Studies in Fuzziness and Soft Computing

Volume 414

The series “Studies in Fuzziness and Soft Computing” contains publications on various topics in the area of soft computing, which include fuzzy sets, rough sets, neural networks, evolutionary computation, probabilistic and evidential reasoning, multi-valued logic, and related fields. The publications within “Studies in Fuzziness and Soft Computing” are primarily monographs and edited volumes. They cover significant recent developments in the field, both of a foundational and applicable character. An important feature of the series is its short publication time and world-wide distribution. This permits a rapid and broad dissemination of research results.

Indexed by SCOPUS, DBLP, WTI Frankfurt eG, zbMATH, SCImago.

All books published in the series are submitted for consideration in Web of Science.

More information about this series at https://link.springer.com/bookseries/2941

Mohammad Yazdi

Linguistic Methods Under Fuzzy Information in System Safety and Reliability Analysis

Springer

Mohammad Yazdi
Faculty of Engineering and Applied Science, Process Engineering
Memorial University of Newfoundland
Newfoundland and Labrador
St. John's, NL, Canada

ISSN 1434-9922 ISSN 1860-0808 (electronic)
Studies in Fuzziness and Soft Computing
ISBN 978-3-030-93351-7 ISBN 978-3-030-93352-4 (eBook)
https://doi.org/10.1007/978-3-030-93352-4

This Springer imprint is published by the registered company Springer Nature Switzerland AG
The registered company address is: Gewerbestrasse 11, 6330 Cham, Switzerland

We express our profound gratitude to all clinicians, nurses, and health service staff around the world and want to dedicate this book in honor of their efforts and selflessly work around the clock to provide care for patients affected with COVID-19.

Preface

System safety and reliability analysis become much more complex over time as different parameters are now required to be considered [1, 2]. Thus, it seems that the conventional methods cannot be that much effective and efficient to improve the system's safety performance. The decision-makers play the most crucial role here, and subjectivity with uncertainty brings many problems to have reliable decision-making in system safety and reliability analysis. Therefore, it needs to understand subjective uncertainty in the system safety and reliability analysis.

The book provides valuable insights into practical linguistic methods under fuzzy information in system safety and reliability analysis and case studies that show the applicability of each proposed approach and programming approach that can be useful for complex systems to quickly implement those to their risk assessment process.

The book is organized to include ten chapters. Chapter 1 discussed the utilization of different linguistic terms reviewed in system safety and reliability analysis. In the case of lack of objective data such as failure rate, loss in dollars, or occurrence probability of an event, the decision-makers are using subjective judgments from experts' opinions in qualitative or linguistic terms. Therefore, industrial sectors need to have an acceptable confidence level for the decisions made based on linguistic terms. In this regard, selecting the proper linguistic terms with a heterogeneous group of decision-makers should be adequately studied. In addition, the inherent feature of each "linguistic term" should be considered for the specific type of decision-making problems like system safety and reliability analysis.

In Chap. 2, the 2-tuple fuzzy-based linguistic term set approach is introduced to analyze system safety and reliability. One of the common linguistic-based approaches is the 2-tuple linguistic model formalized by Herrera and Martnez in 2000. This chapter develops an extension of the 2-tuple linguistic model to analyze system safety and reliability. Besides, the following questions can be justified: (i) what the fuzzy linguistic approach in system safety and reliability is, (ii) how it is preferred among the existing linguistic terms, and (iii) how it can be used in system safety and reliability analysis.

In Chap. 3, the fuzzy bow-tie analysis: concepts, review, and application are discussed. Fault tree (FT) and event tree (ET) analysis methods are the two distinct tools that are often used in process risk assessment. Bow-tie analysis (BTA) as a comprehensive quantitative risk assessment (QRA) model integrates the FT and ET in a common platform to establish a logical connection between causes of an undesirable event and their consequences. To conduct a QRA, it is necessary to estimate the failure probability of basic events and safety barriers. After estimating the severity of the potential consequences, the risk values should be estimated. Experts' judgment elicitation is often employed in QRA due to the scarcity of crisp and precise data. This chapter intends to address an uncertainty aspect (i.e., subjectivity in expert judgment/knowledge) in BTA and applies the fuzzy set theory (FST) to deal with vagueness and imprecision associated with knowledge-based uncertainties. Furthermore, a comprehensive review of the application of FST and fuzzy interface system in the bow-tie model in safety and reliability engineering is presented. To demonstrate the applicability of the fuzzy set theory, a bow-tie diagram of a distillation unit is developed and analyzed by integrating FST into the risk model.

In Chap. 4, optimizing the allocation of risk control measures using the fuzzy MCDM approach in case of review and application is presented. Risk control seeks to eliminate or reduce risk by using various risk control measures (RCMs). Selecting an appropriate set of control measures from a pool of suggested RCMs is challenging work. Three affecting criteria, including risk-related factors (severity, probability, detection rate, and exposure), quality-related factors (applicability, control strategy, cost, duration, reliability, usability), and Haddon matrix components, were recognized. The weight of criteria and sub-criteria on selecting RCMs was calculated using Buckley's fuzzy analytic hierarchy process (FAHP) method. For simplicity, the proposed method does not need a pairwise comparison matrix for RCMs and contains a rating scale with linguistic variables based on the absolute AHP method for each criterion.

In Chap. 5, the fuzzy sets theory and human reliability: review, applications, and contributions are presented. Human reliability analysis (HRA) has drawn increasing attention from academic and industry sectors to enhance system safety in recent decades proactively. HRA mainly focuses on identifying, quantifying, modeling, and preventing human error, which is the most complicated and leading cause of major accidents. However, HRA practitioners have often experienced several severe issues in modeling human behavior owing to rare quantitative data, great uncertainty, and considerable complexity of human behavior. However, few academic attempts have been made to demonstrate how and to what extent HRA has been improved through the FST perspective. Accordingly, this chapter reviews state-of-the-art scientific research to reveal the applications, importance, and contributions of FST onto HRA and its related concerns. It explains the aspects as mentioned earlier by (a) predicting human error probability, (b) quantifying influence of performance shaping factors in human performance, (c) modeling intra-dependency among these factors, (d) incorporating human error into probabilistic safety and risk analysis, (e) modeling human behavior, and finally (f) characterizing the uncertainty analysis in

HRA by incorporating fuzzy set theory. This chapter offers valuable insights into main challenges, gaps, and demands in HRA from both academic and industrial perspectives considering the FST point of view.

In Chap. 6, the emergency decision-making fuzzy-expert-aided disaster management system is studied. The imprecise information and the inaccessibility to data about disasters have been the major factor militating against the efficiency of decision-makers and thus compounding the decision-making process. Robust mathematical approaches are required to respond to disasters quickly and adequately. A new integrated emergency decision-making approach incorporating the best-worst method (BWM), Z numbers, and the zero-sum game is implemented to estimate the importance weights of criteria, the payoffs, and for ranking the various alternative emergency solutions. The efficacy of the proposed approach is illustrated with the Golestan flood disaster of 2019, and an airline emergency relief supplies delivery system is obtained as the optimum solution to the case examined.

In Chap. 7, the smart decision fuzzy-based data envelopment model for failure modes and effects analysis is proposed. Fuzzy-based data envelopment (FDEA) is co-opted with failure mode and effect analysis (FMEA) to form a robust safety and reliability analysis model for the assessment of a system to decipher the potential of failure occurrence. Traditionally, the index for risk premium, which is the risk priority number (RPN) is a product of three risk factors—severity, occurrence, and detection. However, the conventional FMEA methods have been reportedly marred by various shortages. These include RPN scores plagued by ambiguity and vagueness, haphazard scoring and ranking techniques, mode of appraising, evaluating, and inadequate selection of the appropriate corrective actions. This chapter explains and illustrates a new smart decision fuzzy-based data envelopment model that integrates fuzzy set theory, analytical hierarchy process (AHP), and data envelopment analysis (DEA) to handle uncertainty.

In Chap. 8, a probabilistic-based hybrid model (fuzzy-BWM-Bayesian network) is proposed to assess the quality index of medical service. This chapter identified the primary shortages of the typical decision-making methods multi-criteria decision-making (MCDM) like the best–worst method (BWM) highlighted as (i) confidence level, (ii) dynamic feature, and (iii) continues behavior. Then, a probabilistic-based hybrid model is proposed to deal with shortages of MCDM methods. Integrating a BWM with a Bayesian network provides a potential capability for a group of decision-makers to solve a complex decision-making problem in a much more realistic manner. To show the effectiveness of the proposed model, assessing the hospital service quality is studied. The results indicated the advantages of the proposed hybrid method in dealing with shortages of MCDM tools and reflecting the real case approach.

In Chap. 9, a brief review of fuzzy linear programming in system safety and reliability analysis is discussed. As the most effective tool in operations research, Linear programming represents real-world problems using the objective function(s) and constraints to optimize some decision-maker's demands. Fuzzy linear programming (FLP) is when the coefficients in constraints or objective functions or the right-hand-side parameters are fuzzy and vagueness. The FLP problems are categorized into four general classes. This chapter briefly reviews the types of FLP problems, the applications, and some future directions.

This book will be one of the most important guidance books of professionals and researchers working in the field of system safety and reliability. It can be considered as a guide document for an industrial sector to show how they can improve the system safety performance. It also aims to become a valuable reference book for postgraduate and undergraduate students. Finally, as the editor of this book, are grateful to my family, and friends for their constant love, patience, and support. Without their unique support, it would be far beyond the border of my ability and capability to complete this book. With the best wishes that the book will be useful to all concerned.

St. John's, Canada
October 2021

Mohammad Yazdi

References

1. M. Rausand, S. Haugen, *Risk Assessment: Theory, Methods, and Applications* (Wiley, 2020)
2. M. Rausand, A. Hoyland, *System Reliability Theory: Models, Statistical Methods, and Applications*, 664 (2004). https://doi.org/10.1109/WESCON.1996.554026.

Contents

1 A Brief Review of Using Linguistic Terms in System Safety and Reliability Analysis ... 1
Mohammad Yazdi
1.1 The Linguistic Terms in the State of Arts ... 1
1.2 The Linguistic Terms in System Safety and Reliability ... 1
References ... 3

2 2-tuple Fuzzy-Based Linguistic Term Set Approach to Analyse the System Safety and Reliability ... 5
Mohammad Yazdi
2.1 Introduction ... 5
2.2 Background of 2-tuple Linguistic Terms ... 8
2.3 Case Study ... 9
2.3.1 Application Results ... 10
2.4 Conclusions ... 11
References ... 11

3 Fuzzy Bow-Tie Analysis: Concepts, Review, and Application ... 13
Mohsen Omidvar, Esmaeil Zarei, Bahman Ramavandi, and Mohammad Yazdi
3.1 Introduction ... 13
3.2 Sources of Uncertainty in Process Safety Analysis (PSA) ... 16
3.3 Fault Tree Analysis (FTA) ... 18
3.4 Event Tree Analysis (ETA) ... 20
3.5 Bow-Tie Analysis (BTA) ... 22
3.6 Fuzzy Set Theory and Fuzzy Logic ... 26
3.7 Risk Assessment Matrix (RAM) ... 28
3.8 Methodology ... 30
3.8.1 Establishing an expert's Team ... 30
3.8.2 Exploration of Accident Scenarios ... 30
3.8.3 Determination of Fault Tree (FT) and Event Tree (ET) Event Probabilities ... 30

3.8.4 Selection of Fuzzy Membership Function 30
3.8.5 Determination of Linguistic Variables to Define BEs and SFs Probabilities 31
3.8.6 Aggregation of Fuzzy Inputs 31
3.8.7 Transforming Failure Possibility (FPS) to Failure Probability (FP) 33
3.8.8 Determination of Bow-Tie Frequency/probabilities 33
3.8.9 Defuzzification of Aggregated of Fuzzy Values 34
3.8.10 Determination of Consequences Severity 34
3.8.11 Risk Assessment 35
3.9 Case Study 35
3.10 Conclusion 39
References 43

4 Optimizing the Allocation of Risk Control Measures Using Fuzzy MCDM Approach: Review and Application 53
Mostafa Pouyakian, Ashkan Khatabakhsh, Mohammad Yazdi, and Esmaeil Zarei
4.1 Introduction 53
4.2 Typical Framework for Risk Management 53
4.3 Risk Control Models 54
4.3.1 Hierarchy of Control Model 55
4.3.2 Chemical Process Safety Strategies 55
4.3.3 Haddon's Matrix 57
4.4 Evaluation of Risk Control Measures: Conflicting Criteria 58
4.5 Multi-Criteria Decision-Making Methods 59
4.6 Expert Opinions; Vague Linguistic Values 61
4.7 Fuzzy Analytic Hierarchy Process (FAHP) 61
4.7.1 Consistency Ratio 63
4.7.2 Absolute AHP 64
4.8 Application of MCDM Techniques for Safety Optimization Problems 65
4.9 An Application of Fuzzy MCDM for Evaluation of RCMs 66
4.10 A Conceptual Framework for Evaluation of RCMs 66
4.11 General Technique for Evaluating Control Measures (GTECM) 68
4.11.1 Weighting Haddon Matrix with FAHP 69
4.11.2 Weighing Quality and Risk-Related Factors with FAHP 69
4.11.3 Rating RCMs with Absolute AHP 70
4.12 Results of the Overall Score 70
4.13 Classification of the Overall Scores 71
4.14 Application and Results 75
4.15 Conclusion 81
References 86

5 Fuzzy Sets Theory and Human Reliability: Review, Applications, and Contributions 91
Kamran Gholamizadeh, Esmaeil Zarei, Mohsen Omidvar, and Mohammad Yazdi
5.1 Introduction 91
5.2 Human Reliability Analysis Review 93
5.3 Fuzzy Set Contributions and Applications in HRA 94
5.3.1 Predicting Human Error Probability (HEP) 96
5.3.2 Quantifying the Performance Shaping Factors (PSFs) Influence 109
5.3.3 Modeling Intra-dependency Among PSFs 114
5.3.4 HEP in Quantitative Risk Analysis (QRA) 116
5.3.5 Human Behavior (Factor) Model 122
5.3.6 Uncertainty Treatment 128
5.4 Conclusion 131
References 131

6 Emergency Decision Making Fuzzy-Expert Aided Disaster Management System 139
Kehinde Adewale Adesina, Mohammad Yazdi, and Mohsen Omidvar
6.1 Introduction 139
6.2 Methodology 141
6.3 Application of Study 145
6.4 Implications 146
6.5 Conclusion 147
References 147

7 Smart Decision Fuzzy-Based Data Envelopment Model for Failure Modes and Effects Analysis 151
Kehinde Adewale Adesina, Mohammad Yazdi, Esmaeil Zarei, and Mostafa Pouyakian
7.1 Introduction 151
7.2 Material and Methodology 155
7.2.1 Preliminaries of FMEA 156
7.2.2 Expert Judgment 158
7.2.3 Date Envelopment Analysis (DEA) 161
7.3 Application of Study 163
7.4 Conclusion 166
References 167

8 Introducing a Probabilistic-Based Hybrid Model (Fuzzy-BWM-Bayesian Network) to Assess the Quality Index of a Medical Service 171
Mohammad Yazdi, Sidum Adumene, and Esmaeil Zarei
8.1 Introduction 171

8.2 Proposing a New Probabilistic-Based Hybrid Model (Fuzzy-BWM-Bayesian Network) 172
8.3 Application of Study 178
8.4 Conclusion 180
References 181

9 Fuzzy Linear Programming in System Safety 185
Mohammad Yazdi and Arman Nedjati
9.1 Introduction 185
9.2 Classification 186
9.3 Applications of Fuzzy Linear Programming 187
9.3.1 Computer Science Artificial Intelligence 187
9.3.2 Operations Research and Management Science 188
9.3.3 Multi-objective FLP 189
9.4 Future Direction 190
References 190

10 Step Forward on How to Treat Linguistic Terms in Judgment in Failure Probability Estimation 193
Mohammad Yazdi and Esmaeil Zarei
10.1 Introduction 193
10.2 Problem Statement 195
10.3 Conclusion 198
References 198

Chapter 1
A Brief Review of Using Linguistic Terms in System Safety and Reliability Analysis

Mohammad Yazdi

1.1 The Linguistic Terms in the State of Arts

Decision-making in system safety and reliability analysis problem can be generally characterized by an ordered triplet $P = (m, O_m, w_m)$, where the m denotes the number of decision-makers, O_m signifies the opinions from decision-makers m, and the w_m presents the importance weight of decision-makers m. Once all opinions from decision-makers are collected in the specific type of linguistic terms, the qualitative terms will be transferred into the relevant numbers (fuzzy $\rightarrow$ crisp, or crisp). Then, using different aggregation methods and considering the importance weights of the decision-makers, the qualitative terms in numbers are justified into a single value. Figure 1.1 illustrates the history using linguistic terms in a different types of decision-making problems. The detailed discussion regarding linguistic terms under fuzzy information will be explained in the next chapters.

1.2 The Linguistic Terms in System Safety and Reliability

The term "Risk" is the main factor of safety and reliability analysis [19–22], and contains two parameters as the likelihood (L) and consequence (C) [23]. In this regard, the risk is defined as the combination of these two parameters Risk $= L \times C$. This combination provides an answer for two questions (i) what would go wrong, and (ii) what would be the probability of occurrence. Generally, the amount of the risk for an identified should be at an acceptable level or reduced to the tolerable amount. Thus, performing a risk assessment, risk evaluation, and further risk management are the essential tasks for the decision-makers to improve the safety performance of

M. Yazdi (✉)
Faculty of Engineering and Applied Science, Memorial University of Newfoundland, St. John's, NL A1B 3X5, Canada

M. Yazdi, *Linguistic Methods Under Fuzzy Information in System Safety and Reliability Analysis*, Studies in Fuzziness and Soft Computing 414,
https://doi.org/10.1007/978-3-030-93352-4_1

(1) The asymmetric and uniformly distributed linguistic term set	1996
(2) The 2-tuple fuzzy linguistic term set	2000
(3) The virtual linguistic term set	2004
(4) The uncertain linguistic term set	2004
(5) The subscript-symmetric linguistic term set	2004
(6) The proportional 2-tuple fuzzy linguistic term set	2006
(7) The unbalanced linguistic term set	2007
(8) The hesitant fuzzy linguistic term set	2012
(9) The semantic 2-tuple model	2013
(10) The interval-valued hesitant fuzzy linguistic term set	2014
(11) The extended hesitant fuzzy linguistic term set	2015
(12) The generalized asymmetric linguistic term set	2016
(13) The proportional hesitant fuzzy linguistic term set	2016
(14) The probabilistic linguistic term set	2016
(15) The double hierarchy linguistic term set	2017
(16) The double hierarchy hesitant fuzzy linguistic term set	2017
(17) The unbalanced hesitant fuzzy linguistic term set	2019
(18) The Unbalanced double hierarchy linguistic term set	2019

Fig. 1.1 The list of different extensions of linguistic terms set (with references of [1–18])

the system over time. In this book, three main risk assessment tools are considered as FTA (fault tree analysis), ETA (event tree analysis), and FMEA (failure mode and effect analysis).

In FTA, the main objective is deriving the probability of a top event with consideration of causalities from top to down (basic events). The computations of basic events probability can be using logical gates can determine the probability of the top event. In case of a lack of data, decision-makers can express their opinions regarding the probability of the basic event in subjective terms [24–26]. The subjective terms are then transferred into the fuzzy numbers and aggregated in a fuzzy environment. In the end, the fuzzy numbers are defuzzified and can be directly or indirectly be used for the probability of the basic event.

In FMEA, the main objective is deriving the risk priority number for the set of failure modes. Similar to FTA, the group of decision-makers is employed and asked to express their opinions regarding risk factors in FMEA, such as severity, detection, and occurrence. The subjective terms would be transferred into the fuzzy numbers and aggregated in the fuzzy environment. The defuzzification procedure would be applied to reach a crisp value, and this can be used as risk priority numbers.

Such procedures are typically used in system safety and reliability analysis methods. In the following chapter, a couple of main applications and methodologies are explained and studied.

References

1. V. Torra, Negation functions based semantics for ordered linguistic labels. Int. J. Intell. Syst. **11**, 975–988 (1996). https://doi.org/10.1002/(SICI)1098-111X(199611)11:11%3c975::AID-INT5%3e3.0.CO;2-W
2. F. Herrera, L. Martinez, A 2-tuple fuzzy linguistic representation model for computing with words. IEEE Trans. Fuzzy Syst. **8**, 746–752 (2000). https://doi.org/10.1109/91.890332
3. Z. Xu, Deviation measures of linguistic preference relations in group decision making. Omega **33**, 249–254 (2005). https://doi.org/10.1016/j.omega.2004.04.008
4. Z. Xu, Uncertain linguistic aggregation operators based approach to multiple attribute group decision making under uncertain linguistic environment. Inf. Sci. (Ny) **168**, 171–184 (2004). https://doi.org/10.1016/j.ins.2004.02.003
5. J.-H. Wang, J. Hao, A new version of 2-tuple fuzzy linguistic representation model for computing with words. IEEE Trans. Fuzzy Syst. **14**, 435–445 (2006). https://doi.org/10.1109/TFUZZ.2006.876337
6. E. Herrera-Viedma, A.G. López-Herrera, A model of an information retrieval system with unbalanced fuzzy linguistic information. Int. J. Intell. Syst. **22**, 1197–1214 (2007). https://doi.org/10.1002/int.20244
7. R.M. Rodriguez, L. Martinez, F. Herrera, Hesitant fuzzy linguistic term sets for decision making. IEEE Trans. Fuzzy Syst. **20**, 109–119 (2012). https://doi.org/10.1109/TFUZZ.2011.2170076
8. M.-A. Abchir, I. Truck, Towards an extension of the 2-tuple linguistic model to deal with unbalanced linguistic term sets. Kybernetika. **49**, 164–180 (2013). https://www.scopus.com/inward/record.uri?eid=2-s2.0-84876268247&partnerID=40&md5=12c10926a437d7f04eadc33aff32c23d
9. J. Wang, J. Wu, J. Wang, H. Zhang, X. Chen, Interval-valued hesitant fuzzy linguistic sets and their applications in multi-criteria decision-making problems. Inf. Sci. (Ny) **288**, 55–72 (2014). https://doi.org/10.1016/j.ins.2014.07.034
10. H. Wang, Extended hesitant fuzzy linguistic term sets and their aggregation in group decision making. Int. J. Comput. Intell. Syst. **8**, 14–33 (2015). https://doi.org/10.1080/18756891.2014.964010
11. W. Zhou, Z. Xu, Generalized asymmetric linguistic term set and its application to qualitative decision making involving risk appetites. Eur. J. Oper. Res. **254**, 610–621 (2016). https://doi.org/10.1016/j.ejor.2016.04.001
12. Z.-S. Chen, K.-S. Chin, Y.-L. Li, Y. Yang, Proportional hesitant fuzzy linguistic term set for multiple criteria group decision making. Inf. Sci. (Ny) **357**, 61–87 (2016). https://doi.org/10.1016/j.ins.2016.04.006
13. Q. Pang, H. Wang, Z. Xu, Probabilistic linguistic term sets in multi-attribute group decision making. Inf. Sci. (Ny) **369**, 128–143 (2016). https://doi.org/10.1016/j.ins.2016.06.021
14. X. Gou, H. Liao, Z. Xu, F. Herrera, Double hierarchy hesitant fuzzy linguistic term set and MULTIMOORA method: a case of study to evaluate the implementation status of haze controlling measures. Inf. Fusion. **38**, 22–34 (2017). https://doi.org/10.1016/j.inffus.2017.02.008
15. H. Liao, R. Qin, C. Gao, X. Wu, A. Hafezalkotob, F. Herrera, Score-HeDLiSF: a score function of hesitant fuzzy linguistic term set based on hesitant degrees and linguistic scale functions: an application to unbalanced hesitant fuzzy linguistic MULTIMOORA. Inf. Fusion. **48**, 39–54 (2019). https://doi.org/10.1016/j.inffus.2018.08.006
16. H. Wang, J. Yu, H. Fang, H. Wei, X. Wang, Y. Ding, Largely improved mechanical properties of a biodegradable polyurethane elastomer via polylactide stereocomplexation. Polymer (Guildf). **137**, 1–12 (2018). https://doi.org/10.1016/j.polymer.2017.12.067
17. H. Liao, Z. Xu, E. Herrera-Viedma, F. Herrera, Hesitant fuzzy linguistic term set and its application in decision making: a state-of-the-art survey. Int. J. Fuzzy Syst. **20**, 2084–2110 (2018). https://doi.org/10.1007/s40815-017-0432-9

18. Z. Fu, H. Liao, Unbalanced double hierarchy linguistic term set: The TOPSIS method for multi-expert qualitative decision making involving green mine selection. Inf. Fusion. **51**, 271–286 (2019). https://doi.org/10.1016/j.inffus.2019.04.002
19. M. Rausand, S. Haugen, *Risk Assessment: Theory, Methods, and Applications* (Wiley, 2020)
20. N. Khakzad, F. Khan, P. Amyotte, V. Cozzani, Domino effect analysis using Bayesian networks. Risk Anal. **33**, 292–306 (2013). https://doi.org/10.1111/j.1539-6924.2012.01854.x
21. M. Yazdi, Hybrid probabilistic risk assessment using fuzzy FTA and fuzzy AHP in a process industry. J. Fail. Anal. Prev. **17**, 756–764 (2017). https://doi.org/10.1007/s11668-017-0305-4
22. M. Yazdi, F. Khan, R. Abbassi, R. Rusli, Improved DEMATEL methodology for effective safety management decision-making. Saf. Sci. (2020). https://doi.org/10.1016/j.ssci.2020.104705
23. ISO 31000: Risk management. Int. Stand. Organ. (2009)
24. M. Yazdi, S. Kabir, M. Walker, Uncertainty handling in fault tree based risk assessment: State of the art and future perspectives. Process Saf. Environ. Prot. **131** (2019). https://doi.org/10.1016/j.psep.2019.09.003.
25. M. Yazdi, S. Kabir, Fuzzy evidence theory and Bayesian networks for process systems risk analysis. Hum. Ecol. Risk Assess. **26**, 57–86 (2020). https://doi.org/10.1080/10807039.2018.1493679
26. M. Yazdi, F. Nikfar, M. Nasrabadi, Failure probability analysis by employing fuzzy fault tree analysis. Int. J. Syst. Assur. Eng. Manag. **8**, 1177–1193 (2017). https://doi.org/10.1007/s13198-017-0583-y

Chapter 2
2-tuple Fuzzy-Based Linguistic Term Set Approach to Analyse the System Safety and Reliability

Mohammad Yazdi

2.1 Introduction

The original form of fuzzy set theory introduced by Zadeh [1] is highlighting an element could belong to a set of collection. This simply means that there is no partial membership of elements in the set of collections. That is while the fuzzy set theory can be extended within fundamental and concept of classical set like the degree of membership and non-membership in the interval zero and 1.

System safety and reliability analysis, like many practical decision-making problems, are required several aspects of human interventions by processing the rational and cognitive ability of humans [2, 3]. However, there are some circumstances in the information that cannot be quantified due to its features such as "Good", "Bad", and "Excellent". Thus, the decision-makers prefer to express their judgment as much as a natural way rather than numerical value. According to this point, the linguistic terms come up as a notation. In chapter one, the linguistic terms are explained in detail. One of the most common and applicable linguistic terms is developed by Herrera and Martnez [4] in terms of two values known as 2-tuple. The 2-tuple linguistic terms are based on linguistic information and are evaluated in the interval $[-0.5, 0.5)$. The 2-tuple linguistic terms are able to manage the decision-making problem of multi granularity such as [5–8].

The 2-tuple linguistic terms have been comprehensively reviewed in a study by Malhotra and Gupta [9]. Inspired by the preferred reporting items for systematic reviews and meta-analysis (PRISRMA) [10–12], a research analysis is conducted regarding the number of papers published per year in the two decades ending in September 2021. The main application areas of the 2-tuple linguistic terms are specified in the system safety and reliability. The main purpose of this research analysis is to highlight the distribution of the papers in different categories and current research

M. Yazdi (✉)
Faculty of Engineering and Applied Science, Memorial University of Newfoundland, St. John's, NL A1B 3X5, Canada

M. Yazdi, *Linguistic Methods Under Fuzzy Information in System Safety and Reliability Analysis*, Studies in Fuzziness and Soft Computing 414,
https://doi.org/10.1007/978-3-030-93352-4_2

trends in the field of system safety and reliability. This can further provide valuable understanding for researchers working in this area. Firstly, from Fig. 2.1, it can be seen that the 2-tuple linguistic terms and their various improvements have been widely used in different fields.

From Fig. 2.2, it can be observed that the number of publications on the application of 2-tuple linguistic terms in system safety and reliability has increased since the year 2014. This trend considerably increased, especially after the year 2017. It is expected that the studies utilizing the 2-tuple linguistic terms in the system safety and reliability area will continue to grow at an increased pace in the coming decade.

In Table 2.1, some of the highly cited papers are provided by analyzing the "Average Citations per Year" of each publication in the field of safety and risk management. The "Average Citations per Year" means that the number of Web of Science citation indexes for a paper by the end of the year September 2021.

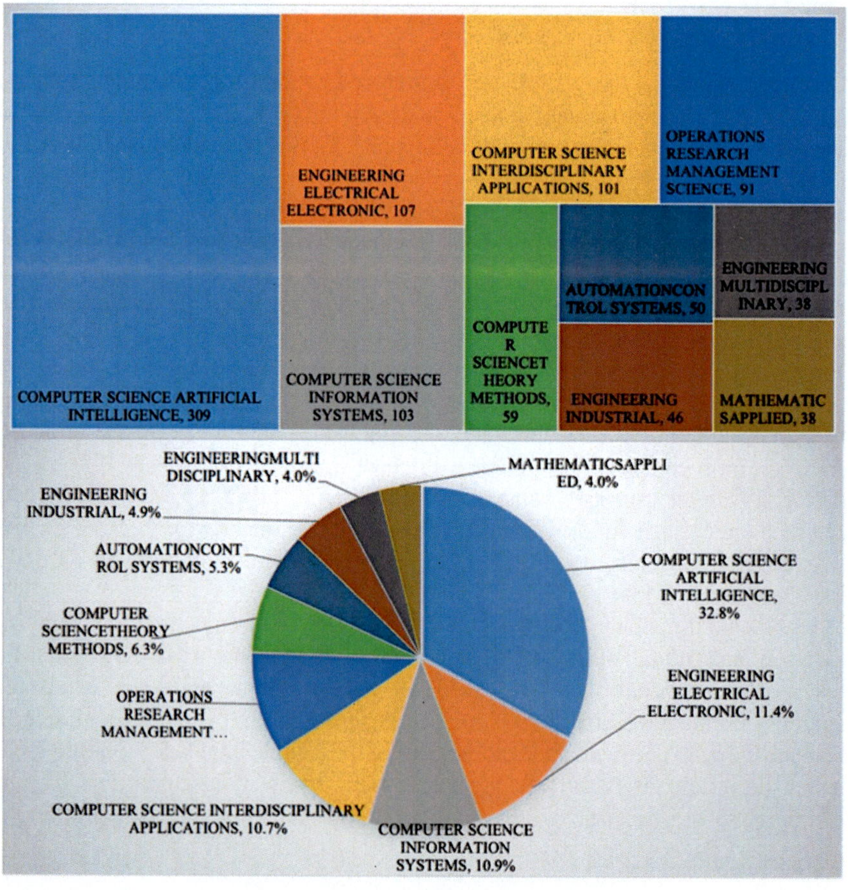

Fig. 2.1 Distribution of published papers "2-tuple linguistic terms" based on applications to September 2021 (*Source* Web of Science)

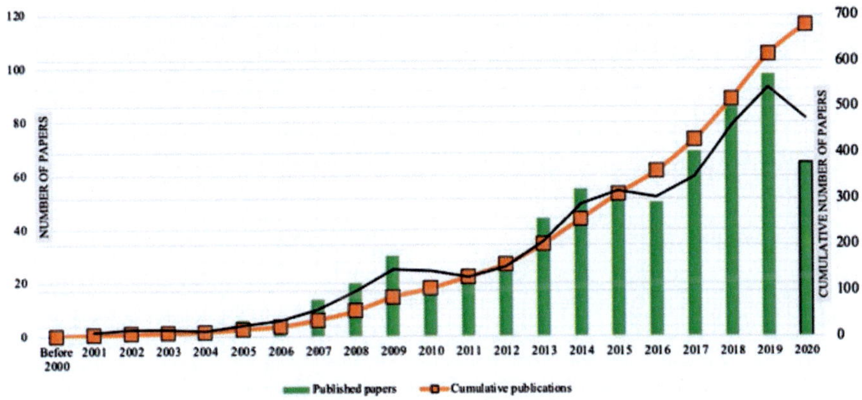

Fig. 2.2 Distribution of published papers "2-tuple linguistic terms" to September 2021 (*Source* Web of Science)

Table 2.1 The highly cited papers based on citation "2-tuple linguistic terms"

Authors	Title	Total citations	Average citation per year
Herrera and Martinez [4]	"A 2-tuple fuzzy linguistic representation model for computing with words"	1642	74.64
Herrera and Martinez [9]	"A model based on linguistic 2-tuples for dealing with multi granular hierarchical linguistic contexts in multi-expert decision-making"	656	31.24
Herrera et al. [8]	"Managing non-homogeneous information in group decision making."	454	26.71
Herrera et al. [6]	"A fuzzy linguistic methodology to deal with unbalanced linguistic term sets."	403	28.79
Wang and Hao [13]	"A new version of 2-tuple. fuzzy linguistic representation model for computing with words."	347	21.69

Source Web of Science

In the following section, a background of 2-tuple linguistic terms underlying the area of system safety and reliability analysis is explained.

2.2 Background of 2-tuple Linguistic Terms

The original concept of 2-tuple linguistic terms is proposed by Herrera and Martinez [4]. In the 2-tuple linguistic terms, the information is in the continuous domains and can be formulated as the following:

Definition 1 Herrera and Martinez [4], let us assume that it is $L = \{l_i | i = 0, \ldots, g\}$ as a finite linguistic term within cardinal order of $g + 1$, and $\beta \in [0, g]$ is a numeric value. Let us consider that $i \in [0, \ldots, g]$ and $\alpha \in [-0.5, 0.5)$ are two values, which is named as "*symbolic translator*". This means that the values denoting the translated value from initial result β into the closest label i. In addition, i represents the round of β, that is, while $\alpha = \beta - i$.

Definition 2 Herrera and Martinez [4], let us assume that it is $L = \{l_i | i = 0, \ldots, g\}$ and defined as a finite linguistic term and let us consider that $\beta \in [0, g]$ is a numeric value denoting the result if a symbolic aggregation process. Therefore, the 2-tuple representing the identical information of β is come up from the following expression:

$$\Delta : [0,\ g] \to \bar{L},$$
$$\Delta(\beta) = \left(l_{round(\beta)}, \beta - i\right)$$

where $\overline{L} = L \times [-0.5, 0.5)$, round(.) is the common round operation by assigning a crisp value to β, an integer value of $i \in [0, \ldots, g]$ closet to β, and $\alpha = \beta - i$ is the term of "*symbolic translation*".

Notice 1: The 2-tuple linguistic term is derived from a linguistic term l_i which includes a value of zero: $l_i \in L \to (l_i, 0) \in L \times [-0.5, 0.5)$.

Notice 2: The linguistic terms according to the 2-tuple should be utilized from any membership functions considering the semantics of given linguistic terms.

Definition 3 Herrera and Martinez [4], Let us assume that there are 2-tuple linguistic variables as (l_i, α_i) and $\left(l_j, \alpha_j\right)$, the following conditions are satisfied:

- If $i < j$, therefore $(l_i, \alpha_i) < \left(l_j, \alpha_j\right)$,
- If $i = j$, therefore:

1. If $\alpha_i = \alpha_j$, therefore $(l_i, \alpha_i) = \left(l_j, \alpha_j\right)$,
2. If $\alpha_i < \alpha_j$, therefore $(l_i, \alpha_i) < \left(l_j, \alpha_j\right)$,
3. If $\alpha_i > \alpha_j$, therefore $(l_i, \alpha_i) > \left(l_j, \alpha_j\right)$.

Definition 4 Herrera and Martinez [4], let us assume that there is $L = \{l_i | i = 0, \ldots, g\}$ as a finite linguistic term within cardinally order of $g + 1$, the negotiation operator can be defied as the following expression:

$$\text{Neg}((l_i, \alpha_i)) = \Delta\big(g - (\Delta^{-1}(l_i, \alpha_i)\big)$$

Definition 5 Tai and Chen [14], let us assume that there is $L = \{l_i | i = 0, \ldots, g\}$ and defined as a finite linguistic term and let us consider that $\beta \in [0, g]$ is a numeric value, therefore, the transformation of β into the 2-tuple linguistic variables is formulated as the following:

$$\Delta : [0, g] \rightarrow \overline{L},$$

$\Delta(\beta) = (l_i, \alpha)$ with two outputs: (i) $\Delta(\beta) = l_i$ when $i = \text{round}(\beta(g))$, and (ii) $\Delta(\beta) = \alpha = \beta - i/g$ where $\alpha \in [-0.5/g, 0.5/g)$.

where $\overline{L} = L \times [-0.5, 0.5)$, round(.) is the general round operation, l_i is the general index label to the β, and α represents the "symbolic translation".

In addition, $\Delta^{-1} : \overline{L} \rightarrow [0, 1]$ can be defined as a conversion of 2-tuple linguistic variables into the equivalent crisp value, $\beta(\beta \in [0, 1])$, by satisfying the following condition:

$$\Delta^{-1}(l_i, \alpha) = \beta = \frac{i}{g} + \alpha.$$

Considering the robustness of 2-tuple linguistic terms as well as several extensions, the 2-tuple linguistic terms have been widely used in the area of system safety and reliability, such as but not limited to [14–18]. In the next section, the 2-tuple linguistic terms are integrated with failure mode and effect analysis (FMEA).

2.3 Case Study

To show that the 2-tuple linguistic terms have enough capability to be used in system safety and reliability, a case study was executed in the aircraft landing system. The FMEA is performed to improve the safety performance of the system. The three different decision-makers (DM1, DM2, and DM3) within different importance weights are involved in the risk assessment process. The employed decision-makers are assigned the relevant importance weights for sharing the expression following the computation procedure of [19, 20]. The obtained importance weights are 0.2, 0.3, and 0.5, respectively. The considered failure modes for risk assessment of aircraft landing systems are listed in Table 2.2.

Table 2.2 The five failure modes of the aircraft landing system

Tag	Failure mode description
FM1	Raising Gear, Fault in raising the wheels
FM2	Coming down Gear, Fault in coming down the wheels
FM3	Fault reporting system, Fault on the run
FM4	Fault reporting system, Wrong run
FM5	Record the result of automatically, Fault on the run

2.3.1 Application Results

In this application section, the five failure modes are taken into account to assess the risk of aircraft landing systems with FMEA [21, 22]. The typical FMEA technique has three risk factors as severity, occurrence, and detection, which the risk priority number (RPN) is defined as the multiplication of the three risk factors [2, 23–25]. The three decision-makers express their judgment based on 2-tuple linguistic terms with reference to [15], and provided in Table 2.3. The total failure modes ranking considering the importance weight of all three decision-makers are provided in Table 2.4.

Table 2.3 The DM1 opinions regrinding the five failure modes in Table 2.2

Tag	Severity	Occurrence	Detection	RPN	Rank
FM1	(S, 0.4)	(S6, 0.1)	(S2, 0.3)	2.94	5
FM2	(S3, 0)	(S6, −0.1)	(S10, −0.3)	4.20	2
FM3	(S6, −0.3)	(S3, −0.4)	(S9, 0.3)	3.79	3
FM4	(S6, −0.2)	(S10, 20.3)	(S4, −0.2)	4.68	1
FM5	(S10, −0.3)	(S2, 0.3)	(S6, −0.4)	3.53	4

Table 2.4 The aggregated RPNs according to the three DMs opinions

Tag	RPN			Aggregated RPN	Rank
	DM1 (0.2)	DM2 (0.3)	DM3 (0.5)		
FM1	2.94	4.54	1.34	2.62	5
FM2	4.20	3.43	4.54	4.139	1
FM3	3.79	2.45	2.45	2.718	4
FM4	4.68	1.45	4.32	3.531	2
FM5	3.53	3.45	3.23	3.356	3

2.4 Conclusions

In this chapter, a common linguistic term as 2-tuple linguistic terms is utilized to improve the FMEA technique. In the existing state of arts, the classical form of FMEA has been reviewed by highlighting the critical shortages [24, 26] such as (i) the relative importance among of risk factors is not considered, (ii) different combination of risk factors can produce the same RPN value, (iii) the risk factors are not easy for decision-makers to be precisely determined, (iv) RPNs are not continuous with many holes, and so on. Using the different linguistic terms like 2-tuple helps decision-makers to deal with the shortages of classical FMEA significantly. In the following Chapters, the most common linguistic terms are reviews, and the utilization of different linguistic terms in system safety and reliability are presented.

References

1. L. Zadeh, Fuzzy sets. Inf. Control. **8**, 338–353 (1965)
2. M. Rausand, S. Haugen, *Risk Assessment: Theory, Methods, and Applications* (Wiley, 2020)
3. T.J. Ross, *Fuzzy Logic with Engineering Applications* (2009). https://doi.org/10.1002/9781119994374
4. F. Herrera, L. Martinez, A 2-tuple fuzzy linguistic representation model for computing with words. IEEE Trans. Fuzzy Syst. **8**, 746–752 (2000). https://doi.org/10.1109/91.890332
5. F. Herrera, S. Alonso, F. Chiclana, E. Herrera-Viedma, Computing with words in decision making: foundations, trends and prospects. Fuzzy Optim. Decis. Mak. **8**, 337–364 (2009). https://doi.org/10.1007/s10700-009-9065-2
6. F. Herrera, E. Herrera-Viedma, L. Martinez, A fuzzy linguistic methodology to deal with unbalanced linguistic term sets. IEEE Trans. Fuzzy Syst. **16**, 354–370 (2008). https://doi.org/10.1109/TFUZZ.2007.896353
7. F. Herrera, E. López, C. Mendaña, M.A. Rodríguez, A linguistic decision model for personnel management solved with a linguistic biobjective genetic algorithm. Fuzzy Sets Syst. **118**, 47–64 (2001). https://doi.org/10.1016/S0165-0114(98)00373-X
8. F. Herrera, L. Martínez, P.J. Sánchez, Managing non-homogeneous information in group decision making, Eur. J. Oper. Res. **166**, 115–132 (2005). https://doi.org/10.1016/j.ejor.2003.11.031
9. T. Malhotra, A. Gupta, A systematic review of developments in the 2-tuple linguistic model and its applications in decision analysis. Soft Comput. (2020). https://doi.org/10.1007/s00500-020-05031-2
10. M. Yazdi, F. Khan, R. Abbassi, R. Rusli, Improved DEMATEL methodology for effective safety management decision-making. Saf. Sci. **127**, 104705 (2020). https://doi.org/10.1016/j.ssci.2020.104705
11. H. Li, J.-Y. Guo, M. Yazdi, A. Nedjati, K.A. Adesina, Supportive emergency decision-making model towards sustainable development with fuzzy expert system. Neural Comput. Appl. (2021). https://doi.org/10.1007/s00521-021-06183-4
12. G.-J. Jiang, H.-X. Chen, H.-H. Sun, M. Yazdi, A. Nedjati, K.A. Adesina, An improved multi-criteria emergency decision-making method in environmental disasters. Soft Comput. (2021). https://doi.org/10.1007/s00500-021-05826-x
13. J.-H. Wang, J. Hao, A new version of 2-tuple fuzzy linguistic representation model for computing with words. IEEE Trans. Fuzzy Syst. **14**, 435–445 (2006). https://doi.org/10.1109/TFUZZ.2006.876337

14. W. Tai, C. Chen, An intellectual capital performance evaluation based on fuzzy linguistic, in: Third Int. Conf. Intell. Inf. Hiding Multimed. Signal Process. (IIH-MSP 2007), (2007), pp. 570–573. https://doi.org/10.1109/IIHMSP.2007.4457774
15. H.-C. Liu, J.-X. You, P. Li, Q. Su, Failure mode and effect analysis under uncertainty: an integrated multiple criteria decision making approach. IEEE Trans. Reliab. **65**, 1380–1392 (2016). https://doi.org/10.1109/TR.2016.2570567
16. H.-C. Liu, P. Li, J.-X. You, Y.-Z. Chen, A novel approach for FMEA: combination of interval 2-tuple linguistic variables and gray relational analysis. Qual. Reliab. Eng. Int. **31**, 761–772 (2015). https://doi.org/10.1002/qre.1633
17. H.-C. Liu, J.-X. You, X.-Y. You, Evaluating the risk of healthcare failure modes using interval 2-tuple hybrid weighted distance measure. Comput. Ind. Eng. **78**, 249–258 (2014). https://doi.org/10.1016/j.cie.2014.07.018
18. C. Mi, Y. Chen, Z. Zhou, C.-T. Lin, Product redesign evaluation: An improved quality function deployment model based on failure modes and effects analysis and 2-tuple linguistic. Adv. Mech. Eng. **10**, 1687814018811227 (2018). https://doi.org/10.1177/1687814018811227
19. M. Yazdi, S. Kabir, A fuzzy Bayesian network approach for risk analysis in process industries. Process Saf. Environ. Prot. **111**, 507–519 (2017). https://doi.org/10.1016/j.psep.2017.08.015
20. M. Yazdi, Hybrid probabilistic risk assessment using fuzzy FTA and fuzzy AHP in a process industry. J. Fail. Anal. Prev. **17**, 756–764 (2017). https://doi.org/10.1007/s11668-017-0305-4
21. M. Yazdi, S. Daneshvar, H. Setareh, An extension to Fuzzy Developed Failure Mode and Effects Analysis (FDFMEA) application for aircraft landing system. Saf. Sci. **98** (2017). https://doi.org/10.1016/j.ssci.2017.06.009
22. S. Daneshvar, M. Yazdi, K.A. Adesina, Fuzzy smart failure modes and effects analysis to improve safety performance of system : Case study of an aircraft landing system. Qual. Reliab. Eng. Int. 1–20 (2020). https://doi.org/10.1002/qre.2607
23. M. Yazdi, Improving failure mode and effect analysis (FMEA) with consideration of uncertainty handling as an interactive approach. Int. J. Interact. Des. Manuf. **13**, 441–458 (2019). https://doi.org/10.1007/s12008-018-0496-2
24. H.C. Liu, FMEA using uncertainty theories and MCDM methods, (2016). https://doi.org/10.1007/978-981-10-1466-6
25. H.-C. Liu, *Improved FMEA Methods for Proactive Healthcare Risk Analysis*, 1st ed., (Springer, 2019). https://doi.org/10.1007/978-981-13-6366-5
26. M. Yazdi, A. Nedjati, E. Zarei, R. Abbassi, A reliable risk analysis approach using an extension of best-worst method based on democratic-autocratic decision-making style, J. Clean. Prod. 120418 (2020). https://doi.org/10.1016/j.jclepro.2020.120418

Chapter 3
Fuzzy Bow-Tie Analysis: Concepts, Review, and Application

Mohsen Omidvar, Esmaeil Zarei, Bahman Ramavandi, and Mohammad Yazdi

3.1 Introduction

Sustainable and dynamic systems require an integrated risk management system (IRMS) to operate in good condition. Risk analysis is a substantial part of RMS. The term "Risk" is a function of three components namely: a set of scenarios (S), the likelihood of events (P), and the consequence of occurrences (C) [1]. This idiom is the cornerstone in the process of decision-making in complex systems [2]. In the risk analysis context, "accident" describes the occurrence of a single event or a chain (sequence) of incidents that can result in undesired consequences [3]. Quantitative Risk Analysis (QRA) is defined as a "systematic approach to predict and decrease the risk of occurrence of an accident" [3–7]. The most important objective of QRA is to collect and consolidate qualitative and quantitative information of potential causes of the undesired events, their consequences, and likelihoods of adverse events [3]. Among existing industries, chemical processes are of great importance from point of view of the accidents occurring within these industries. Due to the handling of huge quantities of hazardous materials, prevention of accidents and other undesired consequences occurring within these industries requires a comprehensive risk assessment

M. Omidvar (✉)
Department of Health, Safety, and Environment (HSE), Faculty of HSEEM, Bushehr University of Medical Sciences, Bushehr, Iran

E. Zarei
Centre for Risk, Integrity and Safety Engineering (C-RISE), Faculty of Engineering and Applied Science, Memorial University of Newfoundland, St. John's, NL A1B 3X5, Canada

B. Ramavandi
Department of Environmental Health Engineering, Faculty of Health and Nutrition, Bushehr University of Medical Sciences, Bushehr, Iran

M. Yazdi
Faculty of Engineering and Applied Science, Memorial University of Newfoundland, St. John's, NL A1B 3X5, Canada

M. Yazdi, *Linguistic Methods Under Fuzzy Information in System Safety and Reliability Analysis*, Studies in Fuzziness and Soft Computing 414,
https://doi.org/10.1007/978-3-030-93352-4_3

process. These unwanted events may consist of injury, sickness, environmental and property damage (fire, explosion, toxic emissions), economic loss or death. Nearly all the process accidents share a common aspect that they initiate from some basic event(s) and result in some final outcomes (consequences). The aim of process safety analysis (PSA) is to predict and analyze probable accident scenario of risk associated with an unintended release of hazardous materials encountered in chemical processes [8].

There are principally three alternatives to accomplish a risk assessment: (1) qualitative (2) semi-quantitative and (3) quantitative [9]. Over the last years, several qualitative and quantitative tools, such as fault tree analysis (FTA), event tree analysis (ETA), hazard and operability study (HAZOP), failure mode and effect analysis (FMEA), have been utilized to perform a risk assessment [10, 11].

QRA techniques follow several systematic steps including, hazard identification, probability assessment, consequence analysis, and risk estimation [12, 13]. The first component encompasses generic qualitative analysis that performs by techniques such as HAZOP, process hazard analysis (PHA), …; while the next items are typical quantitative ones that may be accomplished by methods such as fault tree analysis (FTA), event tree analysis (ETA) [14]. It is worth to note that the common goal of any risk analysis model in PRA is to make sure that a system is constructed and operated to meet allowable risk criteria like ALARP (As Low As Reasonably Practicable) [4, 15]. Each component of the PSA possesses its own characteristic parameters, functions, and models, so entails distinct uncertainty sources associated with the above-mentioned elements [8, 16, 17] fault tree (FT) and event tree (ET) are the two well-known graphical tools applied in PSA to accomplish risk analysis. FT reveals the causes of an unwanted event, while ET demonstrates the consequences (outcomes) of a failure event [2]. These techniques, discretely distinguish the causes and outcomes (consequences) of an undesired event and facilitate the risk and safety assessment through performing a qualitative hazard assessment and a detailed quantitative evaluation of probability [18]. Nevertheless, uncertainties hamper the FTA and ETA in conducting reliable quantitative analyses. A bow-tie (BT) diagram is an integrated concept that incorporates a FT and an ET on the left and right side of the diagram respectively, to demonstrates the risk parameters such as causes, threats and consequences, on a joint platform for mitigating an accident. The left side of the BT diagram demonstrates the pre-events occurrence of accident scenarios, while the right's hand represents the post-events. [6, 19]. The Bow-tie analysis (BTA) has been broadly adopted and applied in multiple industries including oil and gas [2, 20–25], transportation [26, 27], aviation [28], mining [18, 29], chemical process [30, 31], information technology [32–34], and medical and patient safety [35–38].

While in the traditional risk assessment approaches (e.g., FTA, ETA, BTA) the values of the risk parameters are considered to be crisp and deterministic, in the real word these assumptions are not accurate and it is necessary to apply some tools to handle the uncertainties [9, 39, 40]. Some intricacies encountered in QRA include, but are not limited to random variation of components failures and incident consequences, difficulty in acquiring the quantitative information for all components and steps of a process, and interdependencies of various components are not known [5].

Determination, demonstration, and propagation of different uncertainties are significant and essential for BTA since the reliability and credibility of the analysis are intrinsically contingent on the FTA, ETA, as well as BTA [3]. In an attempt to classify the uncertainties associated with PSA, Markowski et al [8] categorized these uncertainties as *the parameter, modeling*, and *completeness* uncertainties. *Parameter uncertainty (PU)* implies to the imprecision and inaccuracies associated with the parameters which are utilized as an input for PSA models. Due to the unavailability or inaccuracy of the data about accidents and inadequate knowledge to inference them, such uncertainties are unavoidable and inherent component of PSA. Inadequacies and imperfection of various models implemented to evaluate probabilities and consequences of the accident scenario introduce the *modeling uncertainty (MU)*. It may be the most important type of uncertainty encountered in consequence assessment. MU is a subjective type of uncertainty arises from knowledge/judgment deduced from experts, which is mostly deficient, imprecise, and defective. The *completeness uncertainty (CU)* is concerned with the consideration of all substantial phenomena and relationships in QRA. While this type of uncertainty is a significant contributor to qualitative hazard analysis, it is difficult to quantify in terms of its magnitude [8]. It is worth noting that, it is not straightforward to put sharp boundaries between these types of uncertainties. Besides, propagation of the above-mentioned uncertainties through each step of PRA and integration of heterogeneous uncertainties to providing an overall estimate of the uncertainty on risk is an intricated process.

Quantitative analysis of a BT is a significant challenge due to the adaption of conventional assumptions in FT and ET analysis. One main assumption is the application of crisp probabilities for the basic events (BEs), safety barriers (SBs), and condition events (CoEs). Assigning the crisp probabilities for these events that usually missing or difficult to be obtained, can introduce data uncertainty in analysis [3]. Data uncertainty is one of the most common and unavoidable uncertainties, generally seen in BTA. In most cases, the probabilities of input events are frequently missing or insufficient and can result in data uncertainty [3, 41]. As the BT applies the conventional operations of FT and ET (conjunction and intersection) for the determination of critical (Top) and output event probabilities, any unaddressed uncertainties in FT and ET finally propagate to the ultimate estimation of CE and OEs quantities in BTA [4].

The risk assessment matrix (RAM) is an instrument that is utilized in different PHA techniques to perform a subjective risk assessment [9]. The foundation of RAM is the definition of risk as a function of severity and frequency of events in a certain accident scenario [1]. Due to the vagueness and incompleteness of the expert's opinion, for determining the probability and severity as well as estimating the risk of different scenarios, it is necessary to use appropriate approaches for handling these uncertainties.

To deal with the data uncertainty in ETA and FTA (i.e., BTA), as well as RAM, expert judgment is frequently used as an alternative to deriving the objective data. This deduced judgment/knowledge may be exposed to some sources of uncertainties including imprecision, subjectivity, vagueness, incompleteness, conflict, and inconsistency [5, 42]. It has been proven that fuzzy set theory (FST) and evidence theory can be exploited to handle the uncertainties encountered in expert-knowledge-based analysis [4]. FST has the potential to handle subjective, imprecise, and vague uncertainty. It can be considered as an extension of traditional set theory (TST) [5, 7]. This theory is an effective tool for applications where no sharp boundaries are possible [15]. FST provides a framework for the translation of qualitative knowledge/beliefs of human beings (i.e., safety experts) into numerical reasoning. To perform risk assessment under uncertainty, FST has been extensively incorporated into different QRA models [43]. These techniques include but not limited to Petri-net (PN) [44, 45], Bayesian network (BN) [19, 46–49], Dynamic Bayesian network (DBN) [20, 50, 51], event tree analysis (ETA) [6, 7, 52, 53], fault tree analysis, (FTA) [54–58], temporal fault trees (TFTs) [59], Bow-Tie [3–5, 25, 60–62], failure mode and effects analysis (FMEA) [63], hazard and operability analysis (HAZOP) [64]. It should be mentioned that there is no universally accepted mathematical process for uncertainty analysis in PRA [15].

Concerning the abovementioned paragraphs, the main objective of this chapter is to develop a comprehensive framework for risk analysis using the BTA method and RAM approach while applying FST to deal with uncertainties. This chapter attempts to incorporate the expert's knowledge to handle missing or incomplete data in BTA and adopts fuzzy set theory and fuzzy logic to addresses the data uncertainty in the QRA context. Besides, a comprehensive review of the application of FST in FTA, ETA, BTA, and RAM is presented. It is worth emphasizing that the current chapter does not aim to address model and quality uncertainties while analyzing the BT diagram. A general framework for PRA using the BTA method has been suggested in Fig. 3.1 that can deal with data uncertainty in risk analysis. A fuzzy-driven approach is developed and exploited to handle the uncertainties due to vagueness, subjectivity, and imprecision assassinated with the expert's knowledge/judgment in BTA. Besides, fuzzy logic is used to extract the expert's knowledge/opinion to determine the risk level through a fuzzy risk assessment matrix. To demonstrate the applicability of the approach in the industrial setting, a BT diagram of a distillation unit accident is developed and analyzed.

3.2 Sources of Uncertainty in Process Safety Analysis (PSA)

Due to the physical variability and knowledge deficiency of the system, uncertainty and hesitancy are intrinsic and inevitable parts of the risk analysis process [4, 19]. Generally, some factors could affect the accuracy of the absolute risk assessment outputs in PSA including, consideration of all significant contributors to the risk magnitude, truthfulness of the mathematical models applied to predict failure specifications and undesired events consequences, and uncertainty arising from the variation of input data (particularly equipment and humans failure data) [15].

Two common traditional approaches for conducting fault tree and event tree analysis, are deterministic and probabilistic methods [6]. The former applies the crisp probability of BEs and estimates the probability of the undesired event (TE) and

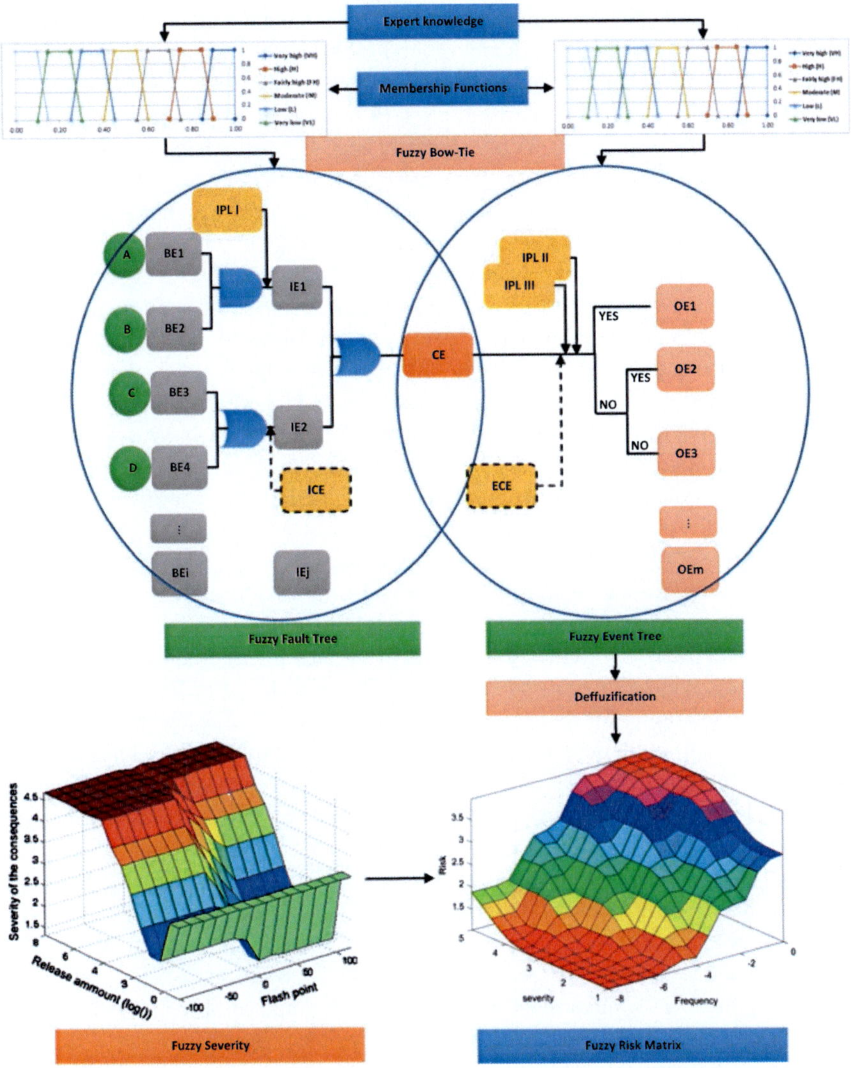

Fig. 3.1 Framework for process risk assessment exploiting the FBTA and FRAM

the frequency/probability of outcome events (OEs) in the FT and ET, respectively. The latter assumes the crisp input's probability as a random variable and characterizes uncertainty utilizing probability density functions (PDF) [7, 65]. One advantage of the deterministic approach is that it provides a quick analysis of FT and ET provided that the probabilities of BEs and SBs are known accurately [7]. One of the most popular approaches that are commonly used in probability theory is the Monte Carlo Simulation (MCS) technique [65–68]. One limitation of the MCS is that requires probability density functions (PDFs) to predict the probability of different outcomes. These PDFs are either obtained from historical data or are acquired by assumption. However, these PDFs are difficult to establish [3, 41, 65]. Besides, MCS entails a large amount of time for simulation while only considers one type of uncertainty (i.e., randomness). Another limitation of MCS is that this technique couldn't consider the knowledge uncertainty especially encountered in PSA (i.e., probability and consequent assessment through FTA and ETA). It is worth noting that knowledge uncertainty is usually more difficult to capture than physical variability [8].

One alternative for conducting QRA in complex systems is to Incorporate expert knowledge/judgments. This approach is particularly advantageous in situations that quantitative data such as the BEs and SFs probabilities are missing or limited [3, 4, 6, 7, 69–72]. Two general types of uncertainty encountered in reliability and risk analysis when using the expert's judgment/knowledge in QRA are aleatory and epistemic uncertainty [3, 15, 42]. The former refers to the inherent variation or random behavior of a system and the surrounding environment. It is also known as, stochastic uncertainty, irreducible uncertainty, and variability. The latter is associated with the lack of information and results from imprecision, indefiniteness, or of vagueness the analyst's knowledge [14]. The abovementioned uncertainties can be originated from three different sources that are known as *quality uncertainty*, *model uncertainty,* and *data uncertainty*. Among these, *quality uncertainty* more often relates to the detailed and comprehensive assessment of hazards, including the identification and explanation of their relationships in the construction of FT and ET. Implementation of some safety and risk evaluation tools such as HAZOP, HAZID, and FMEA can deal with this kind of uncertainty for risk analysis [12, 73, 74]. *Model uncertainty* refers to the mathematical and numerical approximations, model adequacy, and assumptions or validation of the model; while *data uncertainty* associates with the incompleteness, inconsistency (or impreciseness), unavailability, and vagueness (or inadequacy) in input data [75].

Some theories such as probability theory (PT), fuzzy set theory (FST), and evidence theory (or Dempster–Shafer theory, DST) have been proposed to deal with uncertainties in risk analysis [2–5, 7–9, 15, 19, 65, 66, 76, 77]. In an attempt to deal with the data uncertainty in QRA, several surveys have been carried out to apply the expert judgment in the quantification of linguistic statements of the probabilities of the basic events [2, 3, 6–9, 14, 15, 19, 41, 75, 78– 85].

3.3 Fault Tree Analysis (FTA)

Fault tree analysis (FTA) was originally developed in Watson of Bell Labs to analyses the Minuteman launch control system [86]. FTA is a top-down, deductive failure analysis technique that uses Boolean logic to analyze an undesired state of a system [52]. This method demonstrates root (basic) causes that result in an unwanted event (generally an accident) and determines the probability as well as the contribution of basic causes leading to the undesired event [5]. From the risk analysis standpoint, a fault tree (FT) depicts a graphical scheme of the system under consideration through exploring the logical relationship among the causes and occurrence of an unwanted event, typically known as basic events (BEs), and a top event (TE) [87, 88]. Quantitative evaluation of FT is principally dependent on the concept of minimum cut sets (MCSs). MCSs are a series of events leading to the TE. An important element in FTA is the "Gate" concept. There are diffident types of gates in FTA, the most important ones are "And", and "Or" gates [5, 19]. Table 3.2 describes equations and gate operations generally used in traditional FTA and ETA. Considering the independence of BEs, The MCS of TE in the FT can be calculated as follows (Eq. (3.1)) [15]:

$$MCS_{TE(FT)} = \sum_{n=1}^{n} \left(\prod_{i=1}^{i} (BE)_i, \prod_{j=1}^{j} (BE)_j, \ldots, \prod_{k=1}^{k} (BE)_k \right)_n \tag{3.1}$$

The details of FTA can be found elsewhere, but Kumamoto and Henley [89] provided a comprehensive description of FT development and analysis in the process systems.

Two types of inputs for the basic events in FTA are, hardware failure dominated (HFD), and human error dominated (HED) events. Due to the scarcity and imprecision associated with the crisp data for HFD and HED events, it is desirable to use fuzzy numbers to represent these data. The most straightforward approach for HFD events is to consider the relative frequencies of these events as fuzzy numbers. Parameters of the related fuzzy numbers can be obtained from data books or estimated by the "error factor" (EF) concept [90]. In the case of HED events, estimation of the occurrence rate of BEs by a single probability is a difficult approach. In addition, in some cases, only a few or no experimental information are available to obtain creditable estimates of HER. In these cases, the adoption of an expert's opinion in form of linguistic variables can be a beneficial approach. It is worth noting that both HFD and HED event rates can be represented through linguistic variables in situations that there aren't precise, sufficient, and reliable data. [91, 92].

FST has merged with the FTA to relax some uncertainties encountered in FTA (i.e., FFTA). Tanaka et al. [79] used the extension principle to predict the maximum possibility of system failure (TE likelihood) from the possibility of basic events failure (component's failure rate) within the system. Following the Tanaka et al. study, Furuta and Shiraishi [93] applied the FST to determine the state of the TE by the using membership function. They also evaluated the importance of each BE through fuzzy operations. According to Kabir and Papadopoulos [92] four phases to perform the FFTA in a system, include qualitative analysis, fuzzy data approximation, quantitative analysis, and sensitivity analysis. Some outputs of the qualitative analysis phase are the construction of the fault trees, development of fault tree gates and events and determination of minimal cut sets (MCS). Cut sets and fuzzy probabilities of BEs are used in the quantitative analysis phase to derive the fuzzy probabilities of TE. The last phase in FFTA is the sensitivity analysis (SA) of the tree, however, this phase is not always carried out. SA intended to estimate the fuzzy importance measures of the BES, IEs, and other related indexes for specifying the different contributions of the BES, IEs on the occurrence of the TE. The most common importance measures

Table 3.1 Summary of the developments and applications of FCFTA and FDFTA [92, 100]

FTA type	References
Static fuzzy FTA (CFFTA)	Ardeshir et al. [101], Bian et al. [102], Chen et al. [103], Flage et al. [104], Gierczak [105], Kim et al. [106], Kumar and Singh [107], Mhalla et al. [108], Purba [97], Rajakarunakaran et al. [109], Senol and Sahin [110], Tanaka et al. [79], Wang et al. [55–58], Yazdi and Zarei [111], Yazdi et al. [23], Zhang et al. [112] Zonouz and Miremadi [113]
Dynamic fuzzy FTA (DFFTA)	Abdo and Flaus [114], Duan and Fan [115], Duan and Zhou [116], Huang et al. [117], Kabir et al. [59, 118], Tu et al. [119], Yuyan et al. [120]
Applications of FFTA	Aqlan and Ali [60], Duan and Fan [115], Lavasani et al. [53, 95], Gierczak [105], Mhalla et al. [108], Purba et al. [96, 97], Wang et al. [58], Senol et al. [110], Rajakarunakaran et al. [109], Yazdi [111]

for fuzzy inputs in FFTA are the fuzzy importance measure (FIM) [84] and the fuzzy uncertainty importance measure (FUIM) [94]. FFTA has been used in different areas. Yuhua and Datao [78], Wang, Zhang, and Chen [58] and Lavasani et al. [53, 95], have used Fuzzy FTA in process industries, Purba et al. [96] and Purba [97] have applied Fuzzy FTA in nuclear industries. Abdelgawad and Fayek [98] adopted the FFTA to estimate the probability of horizontal directional drilling failure in a pipeline project. Wang et al. [58] used this approach to predict the probability of fire and explosion in a crude oil tank. Lavasani et al. [53] applied the FFTA technique to evaluate the risk of oil and gas leakage in abandoned wells. In another study, Lavasani et al. used this tool to estimate the risk of fire, explosion, and toxic gas releases in the petrochemical industry [95]. Cheliyan and Bhattacharyya [99] exploited the FFTA technique in a subsea production system to calculate the probability of oil and gas equipment leakage. A comprehensive review that demonstrates the developments and applications of FFTA is available in Kabir and Papadopoulos [92] and Mahmood et al. [100] (Table 3.1).

3.4 Event Tree Analysis (ETA)

In contrast to the FTA, Event tree analysis (ETA) is a bottom-up inductive QRA approach that represents the sequence of events in an accident scenario, provided that an initiating event (critical event) has occurred [92]. Event trees visually sketch all sequences that may follow a failure (undesirable consequence) or success (no consequence) [121]. This technique is used to depict the consequences of an undesired event (initiating event) and to approximate the probabilities (frequency) of possible consequences of the initiating event [5]. The initiating event (InE) propagates through a number of occurrences, which are known as interdependent events (IdE). In this sketch, each IdE represents a safety barrier (SB) for escalation of the consequences or

Table 3.2 Equations and operations used in traditional and fuzzy FTA and ETA, as well as BTA [3, 4, 6]

Method	Operation	Traditional	Equation	
			Fuzzy	
FTA	Intersection	$P_{AND} = \prod_{i=1}^{n} P_i$	$\tilde{P}_{AND_F} = \tilde{P}_{X_1} \otimes \tilde{P}_{X_2} \otimes \ldots \otimes \tilde{P}_{X_n}$	(3.3)
			$\tilde{P}_{AND_{TFN}} = \left\{ \prod_{i=1}^{n} l_i, \prod_{i=1}^{n} m_i, \prod_{i=1}^{n} n_i \right\}$	(3.4)
			$\tilde{P}_{AND_{ZFN}} = \left\{ \prod_{i=1}^{n} l_i, \prod_{i=1}^{n} m_i, \prod_{i=1}^{n} n_i, \prod_{i=1}^{n} s_i \right\}$	(3.5)
	Conjunction	$P_{OR} = 1 - \prod_{i=1}^{n} (1 - P_i)$	$\tilde{P}_{OR_F} = 1 - \left[(1 - \tilde{P}_{X_1}) \otimes (1 - \tilde{P}_{X_2}) \otimes \ldots \otimes (1 - \tilde{P}_{X_n}) \right]$	(3.6)
			$P_{OR_{TFN}} = \left\{ 1 - \prod_{i=1}^{n} (1 - l_i), 1 - \prod_{i=1}^{n} (1 - m_i), 1 - \prod_{i=1}^{n} (1 - n_i) \right\}$	(3.7)
			$P_{OR_{ZFN}} = \left\{ 1 - \prod_{i=1}^{n} (1 - l_i), 1 - \prod_{i=1}^{n} (1 - m_i), 1 - \prod_{i=1}^{n} (1 - n_i), 1 - \prod_{i=1}^{n} (1 - s_i) \right\}$	(3.8)
ETA	Intersection	$P_{OE_i} = \lambda \times \prod_{i=1}^{n} P_i$	$\tilde{P}_{OE_i} = \tilde{P}_{CE} \otimes \tilde{P}_{X_1} \otimes \tilde{P}_{X_2} \otimes \ldots \otimes \tilde{P}_{X_n}$	(3.9)

P_i: probability of *i*th (i = 1, 2, 3,..., n) basic events (BEs)
λ: frequency/probability of the initiating event (InE)
P_{OE_i}: frequency/probability of the outcome events (OEi)
$P_{AND_{TFN}}$: the probability of the triangular fuzzy number for AND gate
$P_{AND_{ZFN}}$: the probability of the trapezoidal fuzzy number for AND gate
$P_{OR_{TFN}}$: the probability of the triangular fuzzy number for OR gate
$P_{OR_{ZFN}}$: the probability of the trapezoidal fuzzy number for OR gate

a conditioning event (CoE) [1]. A sequence of interdependent events that commences with an initiating event and terminates in an outcome (OE) establishes a path known as pivotal events (PEs) [122]. ETA can be executed qualitatively or quantitatively. The former identifies the probabilistic events (OEs) that consecutively propagated from an initial event (i.e. TE in BTA) and terminated with the end consequences (or OEs) [123]. The latter approximates the likelihood of incidence of a path (i.e., cut set) by multiplying the probability of an initial event in the probability of all the PEs included in that path [5]. The output at the extremity of each sequence of events is determined by overall MCSs of ET (MCS_{ET}) and can be presented Eq. (3.2) [15]:

$$MCS_{OEs(ET)} = TE.\prod_{m=1}^{m}(IE)_m \tag{3.2}$$

A detailed procedure for developing and analyzing the ETA can be found in AIChE [12] and Lees [124]. As with the FTA, the probabilities of interdependent events (e.g., SBs or CEs) itn ETA may be missing or inaccurately defined. As mentioned earlier, one assumption in traditional FTA and ETA is that the input probabilities of BEs and InE and SBs are known precisely. However, this assumption is often unrealistic and can lead to fallacious conclusions and deviate the goal of risk analysis [15, 75, 125, 126]. Again, FST can be used to deal with the uncertainties associated with these data. In a pioneering study, Kenarangui [82] adopted the fuzzy set logic to consider the imprecision and uncertainty associated with system data while employing ETA. He demonstrated the application of FETA, utilizing a set of event trees in an electric power protection system. Following this study, Huang et al. [90] incorporated fuzzy sets into ETA to account for uncertainties of top events in FTA as initiating events of ETA and human errors. Ramzali et al. [53] applied the FETA technique to analyze the effect of failure in safety barriers in offshore oil and gas drilling systems. Abad and Naeni [52] developed a hybrid framework based on FFTA and FETA to manage changes in construction projects by assessing the change formation.

3.5 Bow-Tie Analysis (BTA)

Bow-tie analysis (BTA) is a consolidated probabilistic tool that investigates accident scenarios in terms of evaluating the likelihoods and pathways of incidents [127]. A bow-tie (BT) diagram incorporates a fault tree (FT) and an event tree (ET) to demonstrate the risk control parameters (physical barriers and administrative controls) on a joint platform to mitigate an accident [3]. The significant concepts encapsulated in BT diagram are preventative barriers and recovery processes. The former barricade a hazardous event (top event) and the latter restrict the escalation of the TE into a more extensive catastrophe [28]. In general, the safety barriers in the BT can be categorized as preventive (IPL I), protective (IPL II), and mitigative (IPL III) layers (Fig. 3.2). It is assumed that preventive and protective layers influence the frequency/probability

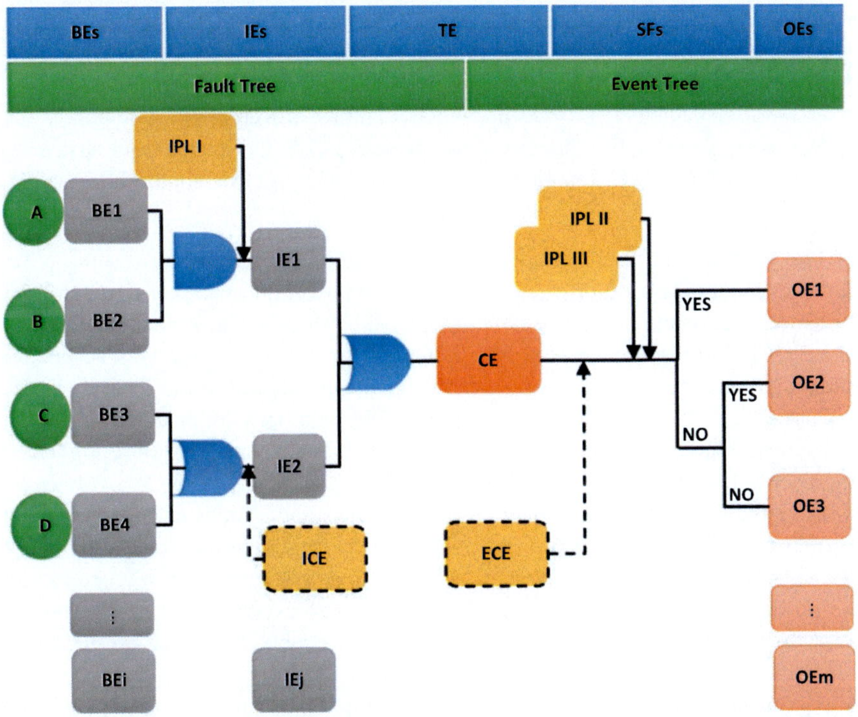

Fig. 3.2 A typical Bow-Tie diagram

of TE and OEs, while the mitigative layer reduces the severity of OEs [19]. Besides, some events can influence the occurrence of CE and OEs internally (e.g., Managerial decisions) or externally (e.g., e.g., ignition, dispersion, wind, presence of the personnel). The former is known as internal conditioning events (ICE) and the latter is known as external conditioning (ECE) events.

The two most important outputs of BT's quantitative analysis are the probabilities of the undesired event (CE) as well as the outcome events (OEs). Figure 3.2 displays five basic elements of a BT and its relationships [3].

- Fault tree (FT): A graphical demonstration of basic events (BEs), intermediate events (IEs), and logic gates (LGs) that represents the sequence of causes leading to the undesired event (Critical event)
- Basic events (causes): These events are the starting point of the FT, and probabilities/frequencies are designated to these events. At the component level, BEs are the fundamental source of failures, faults, malfunctions, or human error. These causes are known as basic events (BEs).
- Critical event (CE): The top event in the FT section of the BT diagram that results from the combination of basic events (BEs). It is also considered as the initiating event (InE) for an ET. This event is equivalent to the TE in the fault tree.

- Event tree (ET): A sequence of possible outcomes of the CE that considers the effect of violating safety functions (e.g., alarm, ESD, Dike) or accident escalation factors (e.g., ignition, dispersion, wind, presence of the personnel, ...) in a dichotomous manner (i.e., yes/no, success/failure, or true/false)
- Outcome events (OE): The last consequences in BT that result from structured propagation of a CE among the safety functions (SFs) or conditioning events (CoEs).

Two approaches for constructing bow-tie diagram are bottom-up and top-down approaches [28]. The top-down approach starts with CE (hazard), which determines the scope and context of the risk assessment [128]. According to Ruijter and Guldenmund [20], three primary types of the bowtie diagram -concerning their respective intentions and requirements- have emerged: an integrated fault tree (FT) an event tree (ET), (i.e., BT), the occupational risk model (ORM), and a detailed diagram emphasizing the direct cause-effect correlations (accident Scenarios). The most applicated approach is the first method (i.e., FT + ET).

Once the BT diagram established, conventional assumptions and mathematical operations of FTA and ETA can be applied to perform quantitative analyses of BT (Table 3.2) [4, 5, 7, 75, 89, 129]. The quantitative assessment of BT determines the probabilities of the top event (CE) and outcome events (OEs) in FT and ET respectively. The main building blocks of the BT as such FT and ET are BEs, CE, and logical gates (AND, OR gates).

An outstanding study that contains the integrated FTA and ETA approach is the ARAMIS project [29, 130, 131]. One of the core elements of ARAMIS was the bowtie diagram. The ARAMIS project yielded the MIMAH methodology (Methodology for the Identification of Major Accident Hazards). Some applications of MIMAH methodology in process industries are, evaluation of emerging risks of carbon capture and storage [132], industrial sites land-use planning (LUP) [133], seaports and offshore terminals operational risk assessment [133], human error analysis of offshore evacuations [134], considering inherent safety for risk-based process plant design [135].

The main objective of the quantitative evaluation of BT is to calculate the probability of the top event (CE) and outcome events (OEs) for FTA or ETA. Assigning the BE's frequency/probability and calculating the TE (or CE) and OEs probabilities is often challenging and extremely coupled with the quality of knowledge concerning the system under consideration and accessibility of accurate data such as likelihood and interdependence of basic (input) events. The accurate probability values of basic events (BEs) are rather rare and are either usually lost, or hard to come by. (Pan and Yun, 1997). To cope with these limitations, FST can be applied in BTA (i.e., FBTA).

Alves et al. inspired an elaborated bowtie diagram on the results of several international databases to assist the management of onshore pipelines operational risks [25]. Suardin et al., developed a screening and conformity tool for fire and explosion assessment (FEA) on Oil and Gas Floating Production Storage Offloading (FPSO) using MS Excel/VBA environment. they incorporated an expert system and applied Bowtie analysis into the tool to describe comprehensive control and recovery measures for the FPSO [62]. Babaleye et al., applied a Bayesian network model based on the bowtie diagram to update the accident data for QRA in the well plugging and abandonment process [47]. Chen et al., exploited the bowtie diagram to construct and simulate a bayesian network (BN) model for analyzing the spatial and temporal variations of the time-dependent parameters in safety of the offshore drilling operations [51]. Stemn et al., applied the BTA to organize and analyze the impressive and effective factors and conditions related to the learning process in safety incidents [61].

Markowski and Kotynia [19] merged LOPA Layer of protection analysis (LOPA) methodology in bow-tie model to increase the benefits of LOPA in the risk management process. Further, they adopted the fuzzy logic system (FLS) to perform a risk assessment of a chemical process. Lu et al. [136] proposed a comprehensive risk evaluation approach by incorporating a risk matrix in a bow-tie model for risk assessment of natural gas pipelines. They also used the FST to calculate the failure probabilities. In a survey, Ferdous et al. [3] attempted to incorporate the expert's knowledge/opinions in bow-tie diagram to handle missing data. They also merged FST and evidence theory to dealt with the uncertainties in bow-tie analysis. Markowski et al. applied fuzzy "IF–THEN" rules in the bow-tie model for explosion risk assessment. They called their methodology fuzzy ExLOPA, that incorporated "IF–THEN" rules in bow-tie construction for explosion risk assessment. Ferdous et al. [5] developed a quantitative risk analysis (QRA) approach, based on the bow tie methodology for risk assessment of "LPG release" and "runaway reaction" scenarios in a process industry. Markowski et al. [15] merged the FT and ET in a bow tie diagram to evaluate the probability of CE and OEs in the scenario of an isobutane storage tank rupture. Shahriar et al. introduced the fuzzy utility value (FUV) notion in bow-tie model to execute the risk assessment of natural gas pipelines employing triple bottom line (TBL) sustainability criteria. Siuta et al. developed a methodology based on the fuzzy bow tie model (FBT), fuzzy risk matrix (FRM), and fuzzy representative loss events selection model (FRLES) for risk assessment in process industries [137]. A detailed review of BT can be found in Ruijter and Guldenmund (Table 3.3) [20].

Table 3.3 Application of Bow-Tie analysis in process industries

Abimbola et al. [138], Aqlan and Mustafa Ali [60], Alves and Lima [25], Badreddine et al. [139–141], Chevreau et al. [130], Deacon et al. [134], de Ruijter and Guldenmund [20], Delvosalle et al. [142], de Dianous and Fiévez [31], Delvosalle et al. [131], Franks et al. [143], IADC [144], Khakzad et al. [145–147]
Ferdous et al. [3–5, 7], Markowski et al. [8, 15, 19, 39], Rostamabadi et al. [24], Shahriar et al. [2], Paltrinieri et al. [148], Rathnayaka et al. [135], Targoutzidis [149], Tugnoli et al. [133], Yazdi, et al. [111], Mokhtari et al. [150], Wilday et al. [132], Zarei et al. [21–23, 71, 151]

3.6 Fuzzy Set Theory and Fuzzy Logic

Fuzzy set theory (FST) was firstly introduced by Zadeh [152]. FST is complementary of classical set theory (CST) that can handle uncertainty and imprecision. While the CST considers membership of an element in a set as a binary value (zero or one), FST allows an element of a set to maintain a membership value between 0–1 [92]. Computing with words (CW) is an alternative to quantal values that can be applied for computing and reasoning through words originated from natural language. The concept of linguistic variables and granulation [153–155] presented in CW, can be used in very complex or vague situations (i.e., those circumstances that are very hard to explain using traditional quantitative expressions) [92].

A fuzzy set A demonstrates a class of objects that encompasses a continuum of grades of membership. This set is distinguished by the membership function, $\mu_A(x)$ which devotes to each object a grade of membership ranging between 0–1. Generally, a fuzzy set can be demonstrated as $A = \{(x, \mu_A(x)); x \in X\}$, where $\mu_A : X \to [0, 1]$ describes the membership function (i.e., degree of belongingness to x in the set A) [15, 92].

Any shape of membership function can be applied to fuzzy sets, depending on the nature of the problem in hand [15]. Typical characteristic function, $\mu_A(x)$, for demonstration of uncertainty in safety and reliability analysis are triangular, trapezoidal, and gaussian convex functions [3, 15, 71, 92]. TFNs and ZFNs are the simplest possible shape of fuzzy numbers that can express the uncertainty in the input probabilities. A TFN can be demonstrated by lower, middle (most likely value), and upper boundary in the form of a vector (pL, pm, pU); while a ZFN is a vector (pl, pm, pn, pu) that can be represented by a lower boundary, left middle, right middle, and upper boundary. Besides, GFNs are used to construct the FLS for the estimation of the risk [4]. After the shape of fuzzy sets is decided, it is necessary to generate the input data in the distinguished shape [92]. Suggested approaches to obtain these inputs are expert knowledge/elicitation, percentage of lower and upper bounds, and 3σ expression [100]. In the current chapter, TFNs and ZFNs are used to quantify subjective and vague uncertainty.

Let x as l, m, n, s $\in \mathbb{R}$. A TFN $\left(\tilde{A}_{TFN} = (l, m, n)\right)$ and a ZFN $\left(\tilde{A}_{ZFN} = (l, m, n, s)\right)$ A could be described by the membership functions μA_{TFN} and μA_{ZFN} respectively, as follows:

$$\mu_{\mathrm{ATFN}}(x) = \begin{cases} \frac{x-l}{m-l} & for a_1 < x < a_2 \\ \frac{n-x}{n-m} & for a_2 \leq x < a_3 \\ 0, & otherwise. \end{cases} \tag{3.10}$$

$$\mu_{\mathrm{A\,ZFN}}(x) = \begin{cases} \frac{x-l}{m-l} & for\ l < x < m \\ 1 & for\ m \leq x \leq n \\ \frac{s-x}{s-n} & for\ n < x < s \\ 0, & otherwise. \end{cases} \tag{3.11}$$

A GFN is specified by two parameters μ and σ; μ demonstrates the membership ship center and σ represents the membership function width. The fuzzy Gaussian function is represented as follows [156].

$$f(x) = ae^{-(x-\mu_f)^2/2\sigma_f^2} \tag{3.12}$$

Mathematical operations of CST (i.e., intersection, union, and complement), can be applied to the FST [15]. The rules of FST [153, 154, 157, 158] and the extension principle [50, 76] are used to perform arithmetic operations for the fuzzy numbers. Fuzzy quantifiers [159] such as, very low (VL), low (L), fairly low (FL), medium (M), fairly high (FH), high (H), and very high (VH) may be used in reliability and safety engineering applications to deal with the "failure possibility" of a component [92].

The notions of FST and FL have been introduced in safety and reliability engineering by Karwowski and Mital in 1986 [160]. In the domain of engineering including safety assessment and risk analysis, FST is now attracted great attention and is recognized as a well-accepted approach, especially concerning vagueness and imprecision handling [3]. The application of FLS in a different area of safety and reliability analysis has been argued in a batch of papers [160, 161]. The foundations of FLS development are already well established [162, 163].

Fuzzy logic (FL) is a concept demonstrated by fuzzy expressions as rules, like: "IF x is A, AND y is B, THEN z is C" where x, y, and z indicate fuzzy variables, and A, B, and C represent fuzzy values. Moreover, statements in the antecedent or descendant of FIS may entail fuzzy logical connectives such as "AND" and "OR" [164]. Fuzzy rule-based modeling (FRBM) is developed as an extension of fuzzy logic (FL). The key difference between these two notions is that the former is used for systems with feedback while the latter is used for no feedback systems [165]. Due to ambiguities associated with the risk components (i.e., probability and severity), the concept of fuzzy logic is suggested in the literature for quantification of the risk. Some stages of the risk assessment process like the construction of risk matrices need to utilize the concepts of fuzzy logic [8, 9, 19]. A fuzzy logic system (FLS) comprises three major components, including:

(1) Fuzzifier: This element maps crisp input into fuzzy sets.
(2) Fuzzy inference system (FIS): The FIS comprises an inference engine that maps input fuzzy sets (IFS), into fuzzy output sets (OFS) employing a set of IF–THEN knowledge rules.
(3) Defuzzifier: this is the last component of an FLS. Defuzzification includes the weighting and averaging the outputs from all discrete fuzzy rules into an individual output value. Defuzzifier delivers the FLS output as a crisp value.

Three types of inference methods that have been commonly used in fuzzy decision making include Mamdani [166], Larsen [167], and Takagi–Sugeno-Kang (TSK) [168]. Among these, the Mamdani inference system (MIS) applies a max–min operator for composition and the min operator as a fuzzy implication. The MIS interface is more compatible with the MFs in the safety and reliability domain and has been broadly implemented in this domain's literature [160, 169]. Recently, the FIS has been advanced by integrating several tools, such as Bayesian reasoning (BR), evidential reasoning (ER) or evidence theory (Dempster–Shafer theory), genetic algorithms (GA), and artificial neural networks (ANNs). The application of these techniques in different domains is summarized in [164].

3.7 Risk Assessment Matrix (RAM)

As mentioned earlier, quantitative risk assessment is accomplished through risk measures assessment. The risk assessment matrix (RAM) is an instrument that is utilized in different PHA techniques to perform a subjective risk assessment (SRA) [9]. The foundation of RAM is the definition of risk as a function of severity and frequency of events in a certain accident scenario [1]. The process of establishing the traditional risk assessment matrix (TRAM) is as follows [9]:

(1) classifying and scaling the frequency/probability and severity of consequences,
(2) classifying and scaling the resulted risk index,
(3) Establishing risk-based rules knowledge,
(4) Presentation of graphical illustration of the risk matrix.

The general type of frequency and severity categorization for application in PRA is the 5 $\times$ 7 cells matrix (e.g., negatable to catastrophic for severity, and remote to very high for likelihood) [170]. Besides, four risk categories are commonly applied in risk and reliability application (i.e., A: acceptable; TA: tolerable-acceptable; TNA: tolerable–unacceptable; and NA: non-acceptable). TRAM is constructed based on risk-based rules knowledge. Risk-based engineering rules (RBER) are utilized to characterize the relevance between different categories of frequency, severity, and risk in TRAM. An application of RBER by using the classical logic implication to exploit the TRM, is demonstrated as follows [9]:

IF frequency is "f" class AND severity is "c" class THEN risk is "r" class.

Due to the vagueness and incompleteness of the expert's opinion, for estimating the risk of different scenarios and assessing the risk level of OEs, the fuzzy risk assessment matrix (FRAM) can be used. FRAM is based on fuzzy logic rules. The output of these rules is described as the FIS. While different MFs can be used to create the FIS, GMF is used here to develop the FRAM. FMs for frequency, severity, and risk levels are shown in Fig. 3.3. As previously explained, severity and probability are selected as the antecedents of FIS, while risk levels are defined as the descendant of FIS. The output of this phase is a fuzzy risk set (FRS) that is obtained by multiplying the event tree outcome probability/frequency (P/F_{OE}) and the highest severity (S) of the consequences. FRS can be converted to the corresponding defuzzified crisp value employing Eq. (3.14) [19]. Figure 3.3 demonstrates the surface of typical FRAM based on a standard rezoning approach [9].

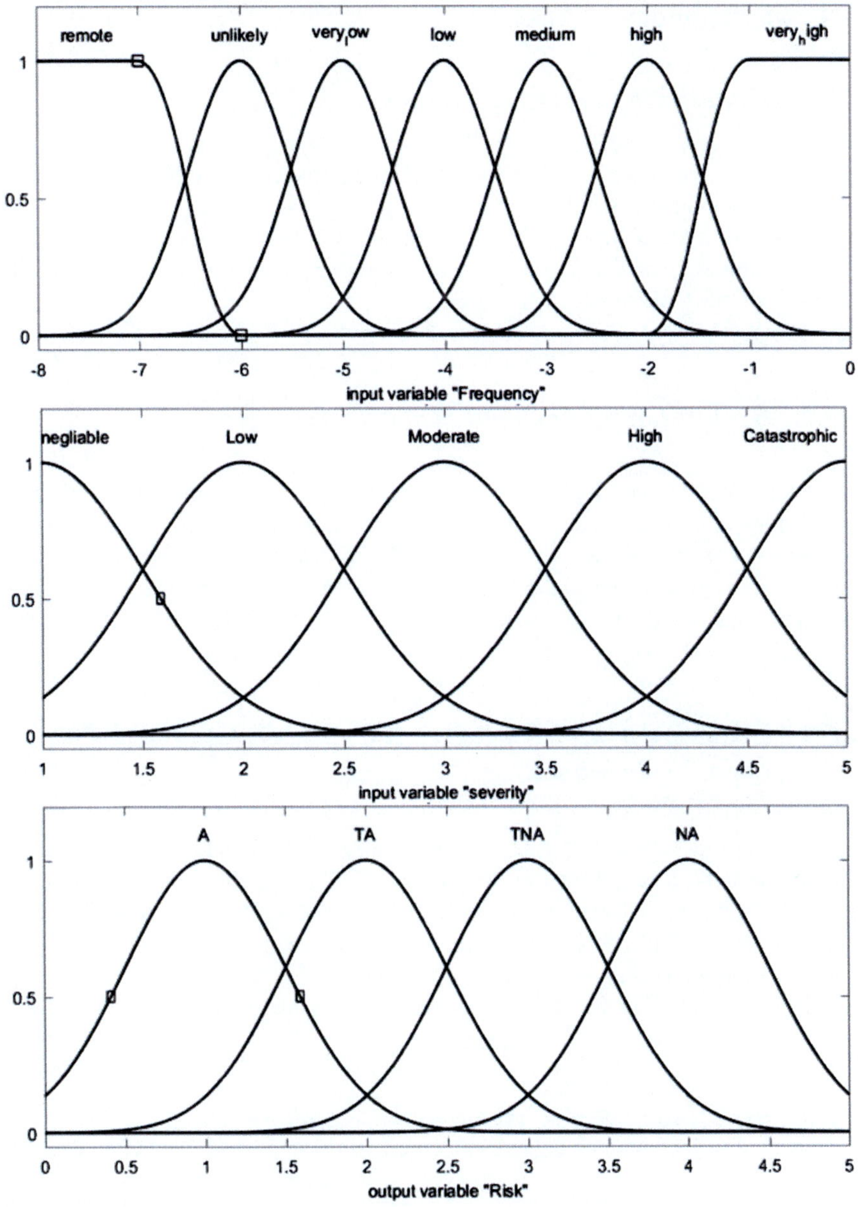

Fig. 3.3 Membership function (MFs) applied for fuzzy risk assessment matrix (FRAM): **a**—frequency MF, **b**—Severity MF, and c- Risk MF

3.8 Methodology

3.8.1 Establishing an expert's Team

Knowledge/judgments elicited from multiple experts always render a preferable approximation than the judgment from a single expert [4]. As the opinion/judgments of the experts in the safety and reliability domain is heterogeneous and Inconsistent, it is advisable to establish a group of experts to represent their opinions about the failure possibility of the BEs, IPLs, and CoEs.

3.8.2 Exploration of Accident Scenarios

The first step of analysis includes the identification of the most important scenarios in the domain of study that can be carried out through different Process Hazard Analysis (PHA) techniques such as HAZOP, FTA, and ETA. The output of this step is the structure of FT and ET (i.e., BT), which can be further analyzed using qualitative and quantitative assessment.

3.8.3 Determination of Fault Tree (FT) and Event Tree (ET) Event Probabilities

Establishing the structure of BT, the minimum cut sets of FT (MCSFT) and ET (MCSET) can be acquired by Boolean algebra [15]. It is worth to note that Eqs. (3.1) and (3.2) can be applied quantitatively only if the frequency/probability of BEs, IPLs, and CoEs are available. As these data are rather scarce and difficult to come by, FST can be used to determine the frequency/probabilities of TE and OEs. The steps of using FST to determine the CE and OEs probabilities are as follows:

3.8.4 Selection of Fuzzy Membership Function

A fuzzy number may be constructed by any normal, convex, and bounded function, while it may be presented through triangular, trapezoidal, and gaussian, singleton shapes. The most important factors for the selection of functions are variable characteristics, achievable information, and expert knowledge/opinion [3]. Because of their simplicity, triangular (TFN), and trapezoidal (ZFN) fuzzy numbers are customarily preferred in reliability and risk analysis [3]. Moreover, Gaussian fuzzy numbers (GFN) may be selected as the most natural and popular alternative for risk analysis i.e., the risk assessment process. [9]. Compared to other fuzzy numbers, a GFN is

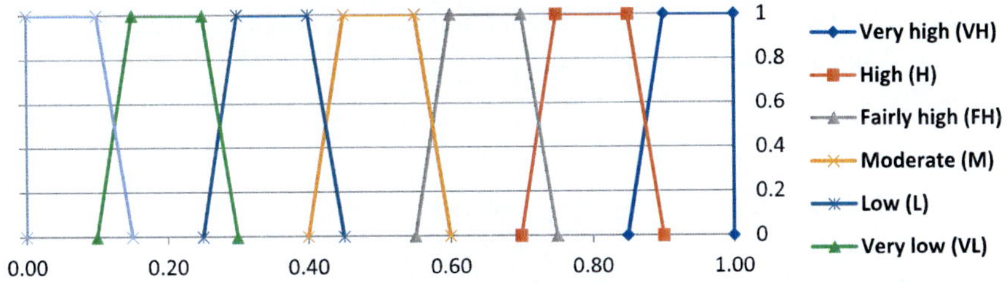

Fig. 3.4 Linguistic variables for the description of BEs, IPLs, and CoEs probabilities

more realistic to depict uncertainty of locus information because the distribution of this membership function is assumed to be Gaussian [156]. Concerning these, ZFN is used to determine CEFP and OEsFP probabilities, while GFN is used to develop a fuzzy risk matrix and to deliver the risk index.

3.8.5 Determination of Linguistic Variables to Define BEs and SFs Probabilities

To determine the CEFP, each expert is requested to represent his/her opinion about the probability of BEs, ICE or SFs (IPL I) in FT section, using the fuzzy scale in Fig. 3.4. This fuzzy scale can be applied to elicit the experts' knowledge/opinions as to the ZFNs for the determination of CEFP and OEsFP.

3.8.6 Aggregation of Fuzzy Inputs

Aggregation renders a mutual agreement and diminishes the conflict among the incompatible sources [4]. Due to the differences in the level of experience, expertise, and background of experts, their opinions/judgments could vary broadly. In order to obtain an agreement among the heterogeneous judgments of the experts, it is essential to merge their opinions [92]. To aggregate fuzzy numbers (judgment /opinions), some generic approaches such as median, mean, min, and max operators are developed [171]. The simplest approach is the arithmetic average of fuzzy numbers (AAFNs). This method considers equal weight for all the experts irrespective of their knowledge, experience, and expertise. In order to take the weightings of the experts/professionals into account, some approaches such as ordered weighted averaging operators [172], coordination index and similarity-based aggregation [173], t-norm and t-conorm [174], Choquet and Sugeno discrete fuzzy integrals [175], symmetric sum [176], iterative optimal aggregation method [177], weighted minimum and the weighted maximum [178], and consistency aggregation method (CAM) [58, 179], similarity aggregation method (SAM) [53, 90, 95, 109] suggested in surveys. The SAM method is adopted here to aggregate the expert's opinion in a single fuzzy number. The details of the SAM method can be found elsewhere [53, 90, 95, 109]. The detailed algorithm of the SAM is described as follows:

3.8.6.1 Determining the Weighting Scores of the Experts

Establishing the of Similarity Matrix (S)

Let M experts provide their opinion/judgment regarding the failure probability of a BEs and SFs in linguistic terms illustrated in Section. The corresponding trapezoidal fuzzy number (ZFN) of expert i th opinion can be written as: $\tilde{A}_i = (a_{i1}, a_{i2}, a_{i3}, a_{i4})$. After all experts expressed their opinion with regard to with each BEs or SFs, the similarity matrix (S) will be established.

$$\mathrm{S} = \begin{bmatrix} 1 & S_{12} & \cdots & S_{1M} \\ S_{21} & 1 & \cdots & S_{2M} \\ \vdots & \vdots & \vdots & \vdots \\ S_{M1} & S_{M2} & \cdots & 1 \end{bmatrix}$$

where S_{ij} denotes the similarity between the judgment of expert i and j (i.e., degree of similarity). It can be determined as:

$$s_{ij} = 1 - \frac{\sum_{K=1}^{4} \left| a_{ik} - a_{jk} \right|}{4}$$

Calculating the average agreement ($AA(Ex_i)$) associated with each expert i as follows:

$$AA(Ex_i) = \frac{\sum_{j=1}^{M} S_{ij}}{M-1}; i \neq j$$

Determining the relative agreement degree ($RAD(Ex_i)$) associated with each expert i as follows:

$$\mathrm{RAD}(Ex_i) = \frac{AA\left(Ex_{ij}\right)}{\sum_{j=1}^{M} AA\left(Ex_{ij}\right)}$$

Calculating the weighting factor ($WF(Ex_i)$) associated with each expert i as follows:

$$\mathrm{WF}(Ex_i) = \frac{WS\left(Ex_{ij}\right)}{\sum_{j=1}^{M} WS\left(Ex_{ij}\right)}$$

Calculating the consensus coefficient ($CC(Ex_i)$), applying weighting factor ($WF(Ex_i)$) and relative agreement degrees ($RAD(Ex_i)$):

$$\mathrm{CC}(Ex_i) = \beta.\mathrm{WF}(Ex_i) + (1 - \beta).\mathrm{RAD}(Ex_i)$$

where β $(0 \leq \beta \leq 1)$ denotes a relaxation factor and is typically considered as $\beta = 0.5$

Calculating the aggregated opinion of the experts as follows:

$$\tilde{A} = \sum_{i=1}^{M} \left(\mathrm{CC}(Ex_i) \times \tilde{A}_i \right)$$

where $\tilde{A}$ is the trapezoidal aggregated opinion of the experts.

3.8.7 *Transforming Failure Possibility (FPS) to Failure Probability (FP)*

Even though probabilities and possibilities, both render a measure of uncertainty and are defined on the interval [0,1], they are rather distinct in nature [180, 181]. The former is related to the degree of belief, likelihood, or frequency whereas the latter is associated with the human sensation of the degree of feasibility [88]. Both probability and possibility notions have been implemented for demonstration of failure data in reliability engineering [75, 81, 84, 96, 182]. The possibility measure could be transformed to the probability rate, to guarantee the compatibility between possibility scores and real numbers [54, 183]. Onisawa [184], described the process of transforming the failure possibility (FPS) to failure probability (FP) and vice versa (Eq. (3.13)).

$$FP = \begin{cases} \frac{1}{10^k}, \; if\ FPS \neq 0 & ; \; k = \left(\frac{1-FPS}{FPS}\right)^{1/3} \times 2.301 \\ 0, & if\ FPS = 0 \end{cases} \tag{3.13}$$

3.8.8 *Determination of Bow-Tie Frequency/probabilities*

After the failure probability of BEs (BEsFP) is elicited, the probabilities of the CE and OEs will be calculated. To determine the fuzzy probability of CE (CEFP), Eqs. (3.3) to (3.8) in Table 3.2 can be used. Once the CEFP is calculated, it will be selected as the initiating event for the ET. subsequently, fuzzy probability of SFs in the ET section of BT (i.e., IPL II and IPL III) and ECE will be extracted from the expert's judgment/knowledge to determine the fuzzy probabilities of the OEs (OEsFP). To perform this, each expert is requested to represent his/her opinion about the probability of ECE and SFs (IPL II and III) in the ET section of BT, using the fuzzy scale in Fig. 3.4. Again, the fuzzy scale in Fig. 3.4 will be used to elicit the failure probability of SFs (in II and III IPL) and ECE and subsequently determining the OEsFP applying the fuzzy arithmetic operations shown in Table 3.2.

3.8.9 Defuzzification of Aggregated of Fuzzy Values

As the final outputs of FFTA and FETA (i.e., CEFP and OEsFP) are in the form of fuzzy numbers, it is necessary to convert the fuzzy numbers into crisp values. Defuzzification is a process of transforming fuzzy numbers into crisp values [92]. Some proposed techniques for defuzzification of fuzzy numbers include the weighted average approach, the centroid approach, the mean max membership approach, the center of maxima approach, the center of area approach, and the mean of maxima approach [185]. The "center of area" (COA) approach is one of the broadly used approaches for the defuzzification process in safety and reliability engineering applications. Using the COA approach, the deffuzified value of a ZFN can be determined as follows [186]:

$$X = \frac{-l_i m_i + n_i s_i + \frac{1}{3}(s_i - n_i)^2 - \frac{1}{3}(m_i - l_i)^2}{-l_i - m_i + n_i + s_i} \tag{3.14}$$

3.8.10 Determination of Consequences Severity

Once the probabilities of the OEs (OEsFP) of BT diagram are calculated, the severity of the consequences will be determined based on look-up tables [187] and fuzzy "if–then" rules. While different parameters can affect the severity of an accident scenario, for the sake of simplicity, only two parameters including potential release amount (PRA) and the type of hazardous substance (e.g., flash point) will be used in this section to construct "if–then" rules of the FIS for severity. Gaussian membership function (GMF) is used to construct the FIS for the determination of severity (Fig. 3.3). As mentioned earlier, the severity of OEs can be mitigated by the IPL III (mitigation layer). Two types of mitigation layers in the ET section are the active systems response time (RT_{AS}) and the emergency response (ER). The effect of these mitigative layers can be introduced as the "severity reduction index" (SRI). Taking into account the effect of SRI, the severity of the OEs can be determined as $S = S_0 - SRI$, and $SRI = MAX(SRI_{AS}, SRI_{ER})$ [17].

Table 3.4 Fuzzy risk assessment matrix (FRAM) variables and their categories

Variable	Linguistic term Category	Range
Frequency (F)	A: Remote B: Unlikely C: Very low D: Low E: Moderate F: High G: Very high	A ($F < 10^{-6}$) B ($10^{-5} \leq F < 10^{-7}$) C ($10^{-4} \leq F < 10^{-6}$) D ($10^{-3} \leq F < 10^{-5}$) E:($10^{-2} \leq F < 10^{-4}$) F ($10^{-1} \leq F < 10^{-3}$) G ($1 < F > 10^{-2}$)
Severity (S)	1: Negligible 2: Low 3: Moderate 4: High 5: Catastrophic	$1 < S \leq 2$ $2 < S \leq 3$ $3 < S \leq 4$ $3 \leq S \leq 5$ $C > S$
Risk (R)	A: acceptable TA: tolerable–acceptable TNA: tolerable–unacceptable NA: unacceptable	$0 < R \leq 2$ $1 \leq R \leq 3$ $2 \leq R \leq 4$ $3 \leq R \leq 5$

3.8.11 Risk Assessment

In the last phase of the risk assessment, the fuzzy risk assessment matrix (RAM) can be used to determine the risk level. The fuzzy logic (FL) concept can be employed to resolve uncertainties (i.e., subjectivity, imprecision) associated with the TRAM. The 5×7 cells risk matrix (i.e., 5 severity level and 7 probability level) was adopted here for risk assessment. Initially, fuzzy set categorization for frequency, severity, and risk levels must be determined (Table 3.4). Subsequently, implementing the fuzzy IF–THEN rules, the FIS of the risk assessment can be stablished (Fig. 3.5).

Having the crisp values of the probability and severity of the undesirable event, the risk value of that can be determined by applying the probability and severity as the inputs of the FIS.

3.9 Case Study

To demonstrate the applicability of the suggested approach in an industrial setting, a BT diagram of a distillation unit accident is developed and analyzed using the FBTA. Distillation columns (DCs) are one of the most important operation units of chemical processes that handle large quantities of flammable liquids. Figure 3.6 demonstrates a typical distillation unit that will be used to analyze its hazards [188].

The selected scenario for risk analysis is the catastrophic loss of containment (LOC) of the column. The fault tree of the LOC is presented in Fig. 3.7. Regarding

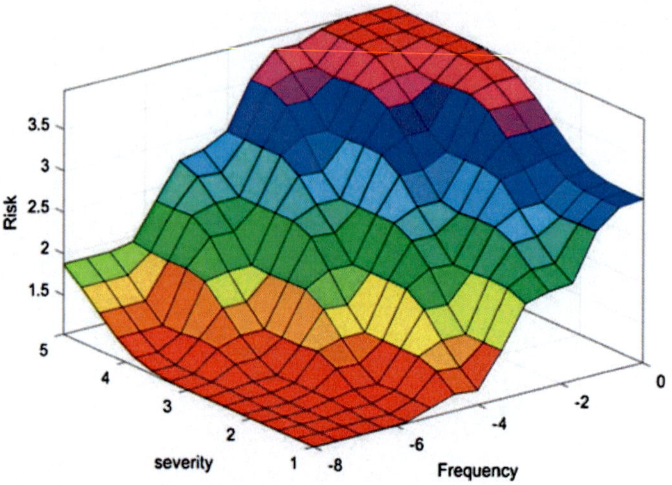

Fig. 3.5 Fuzzy surface (FIS) for risk output

Vílchez et al. [189] the event tree of the LOC is depicted [19]. Figure 3.8 demonstrates the ET of the LOC.

Five experts were selected for this study to represent their opinion/knowledge with regard to the BEs, ICEs, SBs, and ECEs failure probabilities of the BT. Table 3.5 displays the specification of the selected experts. The corresponding weights of the experts are shown in Table 3.5.

For the first step of the study, the experts are requested to express their opinions about the failure possibility of the BEs using the fuzzy scale (ZFNs) shown in Fig. 3.4. Table 3.6 represents the elicited linguistic variables.

After the expert's judgments are captured, the aggregated fuzzy possibility (FPs) of the judgments must be calculated. The last four columns of Table 3.6 demonstrate the aggregated fuzzy probability (AFP) of BEs. Having the AFP of BEs, Eqs. (3.3) to (3.8) in Table 3.2 can be applied to calculate the intermediate events (IEsFP) and critical event (CEFP) fuzzy probabilities of the FT section of BT. Table 3.7 displays the IEsFP and CEFP of the selected scenario.

In the right section of BT (i.e., ET), the fuzzy value of CE (CEFP) is selected as the starting point. Subsequently, the experts are requested to express their opinions about the failure probabilities of SFs (i.e., IPL III) and ECEs using fuzzy Scale in Fig. 3.4. There are two SFs (i.e., automatic deluge system (ADS) and Fire brigade (FB)), two CoEs (i.e., immediate ignition (II) and late ignition (LI)) and a single ECE (i.e., presence of the personnel in the affected area (PP)). Table 3.8 represents the elicited linguistic variables of SFs and ECEs in ET section of BT.

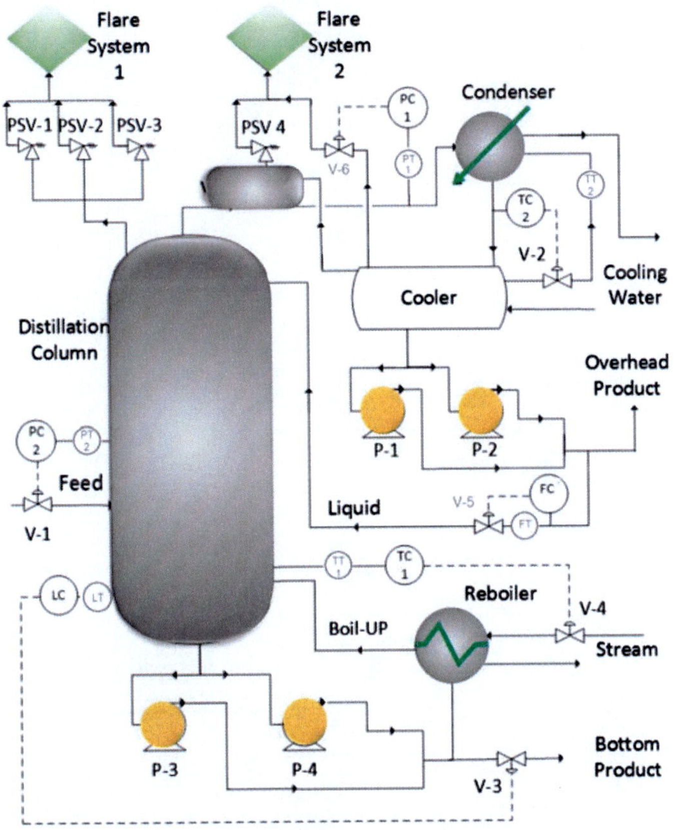

Fig. 3.6 A schematic presentation of the distillation unit

Using the Eq. (3.9), the fuzzy probability of OEs (OEsFP), can be determined. Subsequently, Eq. (3.14) can be used to defuzzify OEsFP. Table 3.9 represents the fuzzy and crisp values of OEs.

Since all OEsFP haven't undesired consequences, only paths that can result in a fatality are considered for determining the total failure probability of the OEs. With regard to this, the total failure probability of the OEs is determined as 3.60E-03. Using capacity and flash point of column's content, the severity (S) value is determined as 3.45. Having the probability (i.e., 3.60E-03) and severity (i.e., 3.45) values, and using the FRAM, the value of risk can be determined as 3.41 (i.e., R = 3.41). Applying the risk surface of the FRAM, it can be shown that this value (R = 3.41) belongs to two risk sets—TNA and NA as is shown in (i.e., TNA: 77%, and NA: 23%).

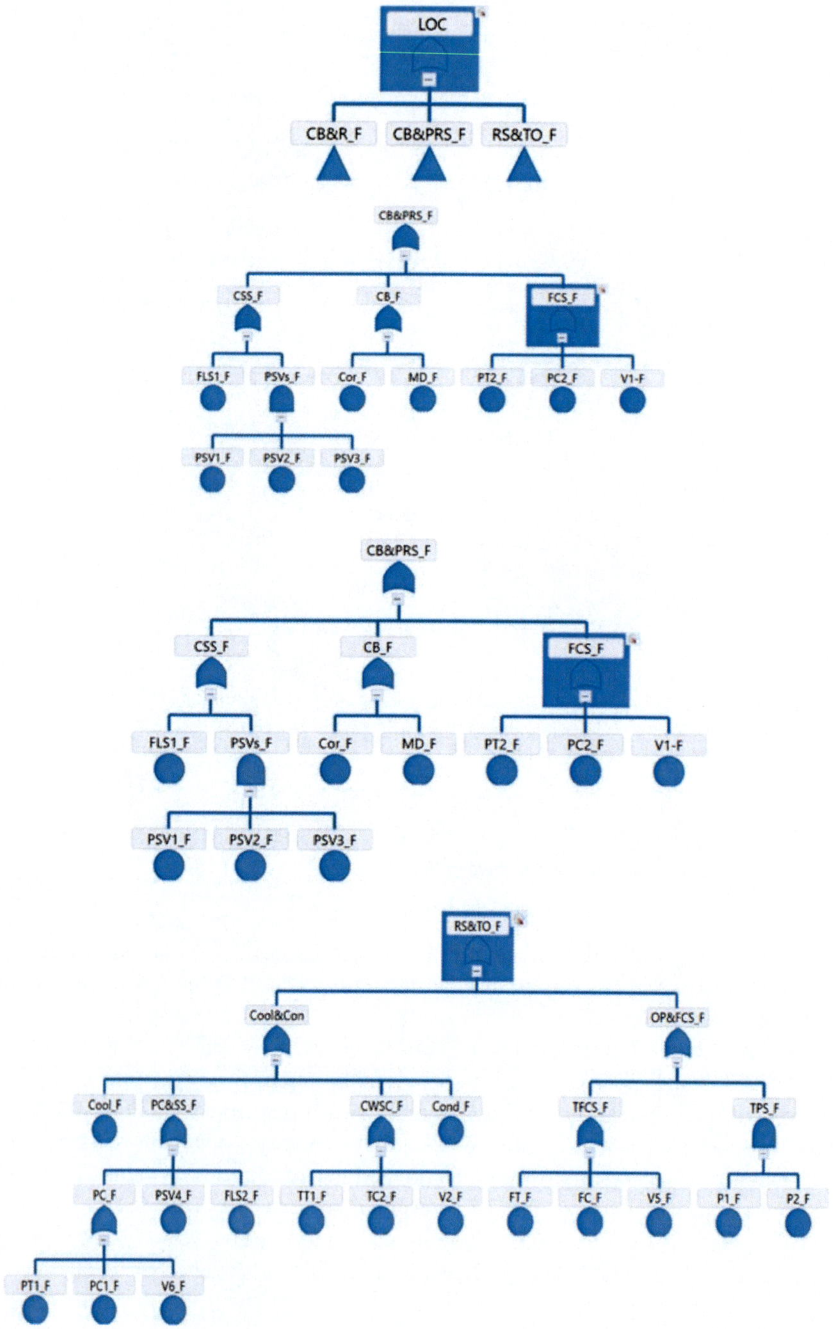

Fig. 3.7 Fault tree of the selected scenario (LOC) of the distillation column

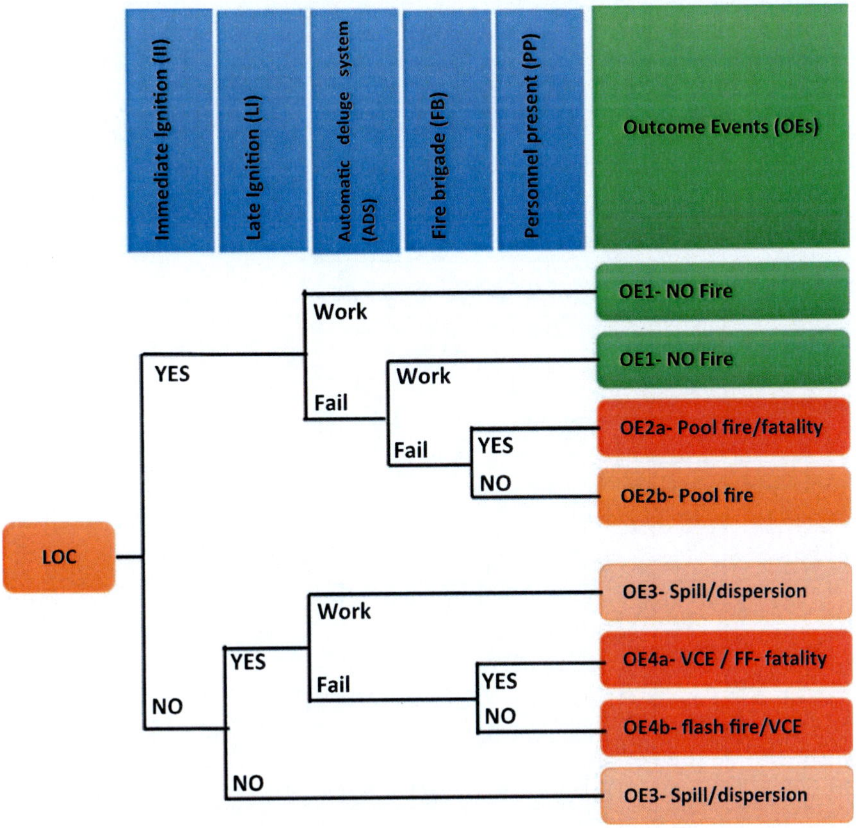

Fig. 3.8 Event tree of the undesired event (LOC)

Table 3.5 Specification of the selected experts

	E1	E2	E3	E4	E5
Title	Senior academic	Engineer	Engineer	Junior academic	Operator
Experience	15	21	18	16	23
Education	PHD	Bachelor	Master	Master	HND
Weighting factor	0.255	0.196	0.196	0.216	0.137

3.10 Conclusion

Risk assessment of process industries involves the investigation of failures modes and consequences of undesirable events in terms of possible losses to assets, personnel, and the environment. Quantifying and characterizing the different risk factors (i.e., probability and severity) connected with the process failures, depends on variables

Table 3.6 Expert's opinion about the failure probability of BEs (BEsFP)

BEs		E1	E2	E3	E4	E5	Fuzzy Probability (FP)			
							L	M	N	U
LC_F	Level control unit failure	VL	L	VL	M	L	2.9E–04	5.9E–04	1.7E–03	2.7E–03
LT_F	Level trainsmiter failure	L	VL	M	VL	L	3.1E–04	6.2E–04	1.8E–03	2.7E03
V3_F	Valve No. 3 failure	VL	VL	L	VL	M	1.6E–04	3.7E–04	1.2E–03	2.0E–03
P3_F	Pump No. 3 failure	L	VL	M	L	L	4.8E–04	8.8E–04	2.3E–03	3.4E–03
P4_F	Pump No. 4 failure	L	VL	M	L	L	4.8E–04	8.8E–04	2.3E–03	3.4E–03
HE_F	Heat exchenger failure	L	VL	M	L	L	4.8E–04	8.8E–04	2.3E–03	3.4E–03
SB_F	Stream blockage failure	M	L	L	VL	L	5.2E–04	9.4E–04	2.4E–03	3.6E–03
TT1_F	Temperature trainsmiter No.1 failure	VL	VL	L	M	VL	1.8E–04	4.1E–04	1.3E–03	2.1E–03
TC1_F	Temperature control unit 1 failure	VL	L	M	L	VL	2.9E–04	6.0E–04	1.7E–03	2.7E–03
V4_F	Valve No. 4 failure	L	VL	L	L	M	4.6E–04	8.6E–04	2.3E–03	3.4E–03
FLS1_F	Flare system no. 1 failure	L	L	M	VL	L	4.7E–04	8.8E–04	2.3E–03	3.4E–03
PSV1_F	Pressure safety valve No. 1 failure	L	M	VL	M	L	7.1E–04	1.2E–03	3.0E–03	4.4E–03
PSV2_F	Pressure safety valve no. 2 failure	L	M	VL	M	L	7.1E–04	1.2E–03	3.0E–03	4.4E–03
PSV3_F	Pressure safety valve no. 3 failure	L	M	VL	M	L	7.1E–04	1.2E–03	3.0E–03	4.4E–03
PT2_F	Pressure Trainsmiter No.2 failure	L	VL	M	FH	L	9.9E–04	1.6E–03	3.7E–03	5.3E–03
PC2_F	Pressure control unit 2 failure	VL	VL	L	FH	M	6.4E–04	1.1E–03	2.8E–03	4.1E–03
V1_F	Valve No. 1 failure	M	L	L	VL	VL	3.6E–04	7.0E–04	1.9E–03	2.9E–03
Cor_F	Corrosion failure	M	L	FH	VL	L	1.0E–03	1.7E–03	3.8E–03	5.5E–03
MD_F	Material defect failure	L	VL	VL	L	M	3.0E–04	6.1E–04	1.8E–03	2.7E–03
Cool_F	Cooling system failure	M	L	FH	L	VL	1.1E–03	1.7E–03	3.9E–03	5.6E–03

(continued)

Table 3.6 (continued)

BEs		E1	E2	E3	E4	E5	Fuzzy Probability (FP)			
							L	M	N	U
LC_F	Level control unit failure	VL	L	VL	M	L	2.9E–04	5.9E–04	1.7E–03	2.7E–03
PT1_F	Pressure trainsmiter no. 1 failure	L	VL	L	M	FH	9.3E–04	1.5E–03	3.6E–03	5.1E–03
PC1_F	Pressure control unit no. 1 failure	VL	VL	L	VL	M	1.6E–04	3.7E–04	1.2E–03	2.0E–03
V6_F	Valve no. 6 failure	VL	L	M	VL	L	2.8E–04	5.7E–04	1.7E–03	2.6E–03
PSV4_F	Pressure safety valve no. 4 failure	L	M	VL	M	L	7.1E–04	1.2E–03	3.0E–03	4.4E–03
FLS2_F	Flare System No. 2 failure	L	L	M	VL	L	4.7E–04	8.8E–04	2.3E–03	3.4E–03
TT2_F	Temperature Trainsmiter No.2 failure	VL	L	L	L	M	4.3E–04	8.0E–04	2.2E–03	3.2E–03
TC2_F	Temperature Control Unit No. 2 failure	L	VL	VL	M	VL	2.1E–04	4.5E–04	1.4E–03	2.2E–03
V2_F	Valve No. 2 failure	VL	L	M	VL	M	4.2E–04	7.9E–04	2.1E–03	3.2E–03
FT_F	Flow Trainsmiter failure	VL	M	L	L	M	6.3E–04	1.1E–03	2.8E–03	4.0E–03
FC_F	Flow Control Logic Unit No. 1 failure	VL	VL	L	L	M	2.7E–04	5.6E–04	1.7E–03	2.6E–03
V5_F	Valve No. 5 failure	L	M	M	VL	VL	4.9E–04	9.0E–04	2.4E–03	3.5E–03
P1_F	Pump No. 1 failure	VL	L	VL	M	VL	1.8E–04	4.2E–04	1.3E–03	2.1E–03
P2_F	Pump No. 2 failure	VL	L	VL	M	VL	1.8E–04	4.2E–04	1.3E–03	2.1E–03
Cond_F	Condenser failure	L	L	VL	M	H	1.2E–03	2.0E–03	4.4E–03	6.2E–03

Table 3.7 Failure probability of IEs (IEsFP) and CE (CEFP) in FFT

IEs	Fuzzy Probability (FP)			
	L	M	N	U
BLC_F	7.59E–04	1.58E–03	4.74E–03	7.35E–03
TCS_F	9.41E–04	1.87E–03	5.33E–03	8.15E–03
HS_F	1.93E–03	3.69E–03	1.00E–02	1.51E–02
BPS_F	2.28E–07	7.79E–07	5.34E–06	1.19E–05
CB&R_F	2.69E–03	5.27E–03	1.47E–02	2.24E–02
PSVs_F	3.60E–10	1.86E–09	2.66E–08	8.26E–08
CSS_F	4.72E–04	8.75E–04	2.30E–03	3.43E–03
CB_F	1.33E–03	2.29E–03	5.58E–03	8.16E–03
FCS_F	1.98E–03	3.44E–03	8.42E–03	1.23E–02
CB&PRS_F	3.77E–03	6.59E–03	1.62E–02	2.37E–02
PC_F	1.37E–03	2.49E–03	6.48E–03	9.69E–03
PC&SS_F	2.55E–03	4.59E–03	1.17E–02	1.74E–02
CWSC_F	1.05E–03	2.05E–03	5.70E–03	8.66E–03
Cool&Condens_F	5.88E–03	1.03E–02	2.55E–02	3.74E–02
TFCS_F	1.39E–03	2.57E–03	6.75E–03	1.01E–02
TPS_F	3.40E–08	1.73E–07	1.80E–06	4.51E–06
OP&FCS_F	1.39E–03	2.57E–03	6.75E–03	1.01E–02
RS&TO_F	7.26E–03	1.28E–02	3.21E–02	4.71E–02
LOC	**1.37E–02**	**2.45E–02**	**6.18E–02**	**9.05E–02**

Bold terms represents in Fig. 3.8

Table 3.8 Expert's opinion about the failure probability of SFs (SFFPs) and CoEs (COEsFP)

	E1	E2	E3	E4	E5	L	M	N	U
II	VL	U	VL	L	VL	0.111	0.151	0.251	0.301
LI	M	M	L	FH	L	0.889	0.849	0.749	0.699
ADS	H	M	FH	FH	M	0.375	0.425	0.525	0.575
FB	H	M	VH	M	L	0.625	0.575	0.475	0.425
PP	L	VL	L	VL	L	0.530	0.580	0.680	0.730

that are often vague, imprecise, and uncertain. FTA and ETA are the two well-recognized tools in safety and reliability engineering. BTA as a comprehensive QRA tools integrates FTA and ETA to deliver the failure probabilities of the CE as well as OEs. An assumption associated with the FTA and ETA (i.e., BTA) is the availability of crisp and precise data when performing QRA. This assumption is not true for all cases and these data may be subjected to some types of uncertainties.

Table 3.9 Failure probability of OEs (OEsFP)

	L	M	N	S	Crisp Values
OE1- No Fire	0.00124	0.00267	0.01040	0.02100	9.16E-03
OE1- No Fire	0.00021	0.00076	0.00324	0.00365	1.96E-03
OE2a- Pool fire/fatality	0.00004	0.00019	0.00111	0.00141	6.91E-04
OE2b- Pool fire	0.00001	0.00008	0.00074	0.00116	5.07E-04
OE3- Spill/dispersion	0.00374	0.00638	0.01631	0.02806	1.39E-02
OE4a- VCE/FF-fatality	0.00061	0.00172	0.00479	0.00457	2.91E-03
OE4b- flash fire/VCE	0.00020	0.00074	0.00320	0.00375	1.97E-03
OE3- Spill/dispersion	0.00760	0.01198	0.02200	0.02691	1.71E-02

An alternative to acquiring the missing data for BEs, SBs, and CoEs is the elicitation of the domain's expert knowledge. But such knowledge elicitation introduces uncertainties and may compromise the credibility of risk analysis. The most important type of uncertainty, especially when exploiting the expert's knowledge/opinion is the impreciseness and vagueness of the extracted data. FST as a well-accepted theory to deal with data uncertainty has attracted great attention in safety and reliability engineering, especially with regard to vagueness and imprecision handling. This chapter introduced an uncertainty aspect in BTA (subjectivity in expert knowledge/judgment) and applied the FST to deal with it. Risk matrix as a handy tool for semi-quantitative risk assessment is also considered advantageous for the determination of risk control measures. Regardless of the usefulness of RAM, some uncertainties can be introduced in different stages of risk estimation (i.e., determination of frequency/probability, severity, and risk index) applying TRM. This chapter also represented the fuzzy logic to handle the uncertainties associated with the risk assessment. Besides, a comprehensive review of the FST and FLS in FTA (FFTA), ETA (FETA), BTA (FBTA), and RAM (FRAM) is presented.

References

1. AIChE. Chemical process quantitative risk analysis. Guidelines Chem. Process Quantitative Risk Anal., pp. 1–55 (2010)
2. A. Shahriar, R. Sadiq, S. Tesfamariam, Risk analysis for oil & gas pipelines: a sustainability assessment approach using fuzzy based bow-tie analysis. J. Loss Prev. Process Ind. **25**(3), 505–523 (2012)
3. R. Ferdous, F. Khan, R. Sadiq, P. Amyotte, B. Veitch, Analyzing system safety and risks under uncertainty using a bow-tie diagram: An innovative approach. Process Saf. Environ. Prot. **91**(1–2), 1–18 (2013)
4. R. Ferdous, F. Khan, R. Sadiq, P. Amyotte, B. Veitch, Handling and updating uncertain information in bow-tie analysis. J. Loss Prevent. Process Industries—J LOSS PREVENT PROC IND. **25** (2012)
5. R. Ferdous, F. Khan, R. Sadiq, P. Amyotte, B. Veitch, Fault and event tree analyses for process systems risk analysis: uncertainty handling formulations. Risk Anal. **31**(1), 86–107 (2011)

6. R. Ferdous, F. Khan, R. Sadiq, P. Amyotte, B. Veitch, Fault and event tree analyses for process systems risk analysis: uncertainty handling formulations. Risk Anal.: Int. J. **31**(1), 86–107 (2011)
7. R. Ferdous, F. Khan, R. Sadiq, P. Amyotte, B. Veitch, Handling data uncertainties in event tree analysis. Process Saf. Environ. Prot. **87**(5), 283–292 (2009)
8. A.S. Markowski, M.S. Mannan, A. Kotynia, D. Siuta, Uncertainty aspects in process safety analysis. J. Loss Prev. Process Ind. **23**(3), 446–454 (2010)
9. A.S. Markowski, M.S. Mannan, Fuzzy risk matrix. J. Hazard. Mater. **159**(1), 152–157 (2008)
10. F.I. Khan, S. Abbasi, Risk analysis of a typical chemical industry using ORA procedure. J. Loss Prev. Process Ind. **14**(1), 43–59 (2001)
11. E. Zarei, M.J. Jafari, N. Badri, Risk assessment of vapor cloud explosions in a hydrogen production facility with consequence modeling. J. Res. Health Sci. **13**(2), 181–187 (2013)
12. AIChE. Guidelines for chemical process quantitative risk analysis: center for Chemical Process Safety of the American Institute of Chemical (2000)
13. A. Dormohammadi, E. Zarei, M.B. Delkhosh, A. Gholami, Risk analysis by means of a QRA approach on a LPG cylinder filling installation. Process Saf. Prog. **33**(1), 77–84 (2014)
14. D. Siuta, A.S. Markowski, M.S. Mannan, Uncertainty techniques in liquefied natural gas (LNG) dispersion calculations. J. Loss Prev. Process Ind. **26**(3), 418–426 (2013)
15. A.S. Markowski, M.S. Mannan, A. Bigoszewska, Fuzzy logic for process safety analysis. J. Loss Prev. Process Ind. **22**(6), 695–702 (2009)
16. M. Yazdi, S. Kabir, M. Walker, Uncertainty handling in fault tree based risk assessment. State of the art and future perspectives.Process Saf. Environ. Prot. **131**, (2019). https://doi.org/10.1016/j.psep.2019.09.003
17 M. Yazdi, N.A. Golilarz, K.A. Adesina, A. Nedjati, Probabilistic Risk Analysis of Process Systems Considering Epistemic and Aleatory Uncertainties. A Comparison Study. Int. J. Uncertainty, Fuzziness Knowledge-Based Syst. **29**, 181–207 (2021). https://doi.org/10.1142/S0218488521500098
18. J. Spouge, A guide to quantitative risk assessment for offshore installations: CMPT Aberdeen (1999)
19. A.S. Markowski, A. Kotynia, "Bow-tie" model in layer of protection analysis. Process Saf. Environ. Prot. **89**(4), 205–213 (2011)
20. A. de Ruijter, F. Guldenmund, The bowtie method: a review. Saf. Sci. **88**, 211–218 (2016)
21. E. Zarei, N. Khakzad, V. Cozzani, G. Reniers, Safety analysis of process systems using Fuzzy Bayesian Network (FBN). J. Loss Prev. Process Ind. **57**, 7–16 (2019)
22. E. Zarei, A. Azadeh, N. Khakzad, M.M. Aliabadi, I. Mohammadfam, Dynamic safety assessment of natural gas stations using Bayesian network. J. Hazard. Mater. **321**, 830–840 (2017)
23. E. Zarei, M. Yazdi, R. Abbassi, F. Khan, A hybrid model for human factor analysis in process accidents: FBN-HFACS. J. Loss Prev. Process Ind. **57**, 142–155 (2019)
24. A. Rostamabadi, M. Jahangiri, E. Zarei, M. Kamalinia, M. Alimohammadlou, A novel Fuzzy Bayesian Network approach for safety analysis of process systems; an application of HFACS and SHIPP methodology. J. Clean. Prod. **244**, 118761 (2020)
25. D.T.S. Alves, G.B.A. Lima, Establishing an onshore pipeline incident database to support operational risk management in Brazil - Part 2: Bowtie proposition and statistics of failure. Process Saf. Environ. Prot. **155**, 80–97 (2021)
26. F. Delmotte, A sociotechnical framework for the integration of human and organizational factors in project management and risk analysis. Virginia Tech (2003)
27. K. Van Scyoc, G. Hughes, Rail ruminations for process safety improvement. J. Loss Prev. Process Ind. **22**(6), 689–694 (2009)
28. J. Aust, D. Pons, A systematic methodology for developing bowtie in risk assessment: application to borescope inspection. Aerospace **7**(7), 86 (2020)
29. R. Burgess-Limerick, T. Horberry, L. Steiner, Bow-tie analysis of a fatal underground coal mine collision. Ergonomics Australia. **10**(2) (2014)

30. I. Papazoglou, L. Bellamy, A. Hale, O. Aneziris, B. Ale, J. Post et al., I-Risk: development of an integrated technical and management risk methodology for chemical installations. J. Loss Prev. Process Ind. **16**(6), 575–591 (2003)
31. V. De Dianous, C. Fievez, ARAMIS project: a more explicit demonstration of risk control through the use of bow–tie diagrams and the evaluation of safety barrier performance. J. Hazard. Mater. **130**(3), 220–233 (2006)
32. H. Abdo, M. Kaouk, J.-M. Flaus, F. Masse, A safety/security risk analysis approach of industrial control systems: a cyber bowtie–combining new version of attack tree with bowtie analysis. Comput. Secur. **72**, 175–195 (2018)
33. K. Bernsmed, C. Frøystad, P. H. Meland, D. A. Nesheim, Ø.J. Rødseth (eds). Visualizing cyber security risks with bow-tie diagrams. International Workshop on Graphical Models for Security (Springer, 2017)
34. A. Faulkner, M. Nicholson, *Data-Centric Safety: Challenges, Approaches, and Incident Investigation* (Elsevier, 2020)
35. M. Culwick, A. Merry, D. Clarke, K. Taraporewalla, N. Gibbs, Bow-tie diagrams for risk management in anaesthesia. Anaesth. Intensive Care **44**(6), 712–718 (2016)
36. Z. Abdi, H. Ravaghi, M. Abbasi, B. Delgoshaei, S. Esfandiari, Application of Bow-tie methodology to improve patient safety. Int J. Health Care Qual. Assur. (2016)
37. M.C. Kerckhoffs, A.F. van der Sluijs, J.M. Binnekade, D.A. Dongelmans, Improving patient safety in the ICU by prospective identification of missing safety barriers using the bow-tie prospective risk analysis model. J. Patient Saf. **9**(3), 154–159 (2013)
38. H. Elamir, Enterprise risk management and bow ties: going beyond patient safety. Bus. Process Manage. J. (2019)
39. M. Yazdi, Introducing a heuristic approach to enhance the reliability of system safety assessment. Qual. Reliab. Eng. Int. **35**, (2019). https://doi.org/10.1002/qre.2545
40. S. Kabir, M. Yazdi, J.I. Aizpurua, Y. Papadopoulos, Uncertainty-Aware Dynamic Reliability Analysis Framework for Complex Systems. IEEE Access **6**, (2018). https://doi.org/10.1109/ACCESS.2018.2843166
41. A. S. Markowski, D. Siuta, Application of fuzzy logic approach to consequence modeling in process industries. Chem. Eng. Trans. (2013)
42. Ayyub BM, Klir GJ. Uncertainty modeling and analysis in engineering and the sciences: Chapman and Hall/CRC; 2006.
43. X. Guo, J. Ji, F. Khan, L. Ding, Q. Tong, A novel fuzzy dynamic Bayesian network for dynamic risk assessment and uncertainty propagation quantification in uncertainty environment. Safety Sci. **141**, 105285 (2021)
44. N.K. Jyotish, L.K. Singh, C. Kumar, A state-of-the-art review on performance measurement petri net models for safety critical systems of NPP. Ann. Nuclear Energy **165**, 108635 (2022)
45. R. Squillante Jr., J.C. Dias, L.A. Moscato, F. Junqueira, P.E. Miyagi, D.J.D. Santos Filho, A framework for synthesis of safety-related control design to avoid critical faults and pathogenic accidents in the process industries. Safety Sci. **139**, 105168 (2021)
46. L. Zhang, X. Wu, Y. Qin, M.J. Skibniewski, W. Liu, Towards a fuzzy Bayesian network based approach for safety risk analysis of tunnel-induced pipeline damage. Risk Anal. **36**(2), 278–301 (2016)
47. A.O. Babaleye, R.E. Kurt, F. Khan, Safety analysis of plugging and abandonment of oil and gas wells in uncertain conditions with limited data. Reliab. Eng. Syst. Saf. **188**, 133–141 (2019)
48. M. Pouyakian, M.J. Jafari, F. Laal, F. Nourai, E. Zarei, A comprehensive approach to analyze the risk of floating roof storage tanks. Process Saf. Environ. Prot. **146**, 811–836 (2021)
49. E. Zarei, A. Karimi, E. Habibi, A. Barkhordari, G. Reniers, Dynamic occupational accidents modeling using dynamic hybrid Bayesian confirmatory factor analysis: An in-depth psychometrics study. Safety Sci. **136**, 105146 (2021)
50. J. Yao, J. Li, H. Li, X. Wang, (eds). Modeling system based on fuzzy dynamic Bayesian network for fault diagnosis and reliability prediction. 2015 Annual Reliability and Maintainability Symposium (RAMS), IEEE (2015)

51. K. Chen, X. Wei, H. Li, H. Lin, F. Khan, Operational risk analysis of blowout scenario in offshore drilling operation. Process Saf. Environ. Prot. **149**, 422–431 (2021)
52. F. Abad, L.M. Naeni, A hybrid framework to assess the risk of change in construction projects using fuzzy fault tree and fuzzy event tree analysis. Int. J. Construct. Manage., 1–13 (2020)
53. S.M. Lavasani, N. Ramzali, F. Sabzalipour, E. Akyuz, Utilisation of fuzzy fault tree analysis (FFTA) for quantified risk analysis of leakage in abandoned oil and natural-gas wells. Ocean Eng. **108**, 729–737 (2015)
54. L. Shi, J. Shuai, K. Xu, Fuzzy fault tree assessment based on improved AHP for fire and explosion accidents for steel oil storage tanks. J. Hazard. Mater. **278**, 529–538 (2014)
55. J. Wang, R. Nie, H.Y. Zhang, X.H. Chen, New operators on triangular intuitionistic fuzzy numbers and their applications in system fault analysis. Inf. Sci. **251**, 79–95 (2013)
56. H. Wang, X. Lu, Y. Du, C. Zhang, R. Sadiq, Y. Deng, Fault tree analysis based on TOPSIS and triangular fuzzy number. Int. J. Syst. Assur. Eng. Manage. **8**(4), 2064–2070 (2017)
57. D. Wang, Y. Zhang, X. Jia, P. Jiang, B. Guo, Handling uncertainties in fault tree analysis by a hybrid probabilistic–possibilistic framework. Qual. Reliab. Eng. Int. **32**(3), 1137–1148 (2016)
58. D. Wang, P. Zhang, L. Chen, Fuzzy fault tree analysis for fire and explosion of crude oil tanks. J. Loss Prev. Process Ind. **26**(6), 1390–1398 (2013)
59. S. Kabir, M. Walker, Y. Papadopoulos, E. Rüde, P. Securius, Fuzzy temporal fault tree analysis of dynamic systems. Int. J. Approximate Reason. **77**, 20–37 (2016)
60. F. Aqlan, E.M. Ali, Integrating lean principles and fuzzy bow-tie analysis for risk assessment in chemical industry. J. Loss Prev. Process Ind. **29**, 39–48 (2014)
61. E. Stemn, C. Bofinger, D. Cliff, M.E. Hassall, Failure to learn from safety incidents: status, challenges and opportunities. Saf. Sci. **101**, 313–325 (2018)
62. J.A. Suardin, A. Jeff McPhate, A. Sipkema, M. Childs, M.S. Mannan, Fire and explosion assessment on oil and gas floating production storage offloading (FPSO): an effective screening and comparison tool. Process Saf. Environ. Prot. **87**(3), 147–160 (2009)
63. S. Hassan, J. Wang, C. Kontovas, M. Bashir, Modified FMEA hazard identification for cross-country petroleum pipeline using Fuzzy Rule Base and approximate reasoning. J. Loss Prevent. Process Industries. **74**, 104616 (2022)
64. M. Cheraghi, A.E. Baladeh, N. Khakzad, A fuzzy multi-attribute HAZOP technique (FMA-HAZOP): application to gas wellhead facilities. Saf. Sci. **114**, 12–22 (2019)
65. R.C. Wilcox, B. M. Ayyub, (eds), Uncertainty modeling of data and uncertainty propagation for risk studies. Uncertainty Modeling and Analysis, International Symposium on 2003, IEEE Computer Society
66. M. Abrahamsson, *Uncertainty in Quantitative Risk Analysis-Characterisation and Methods osf Treatment*, Lund University (2002)
67. K. Park, G. Lee, C. Kim, J. Kim, K. Rhie, W.B. Lee, Comprehensive framework for underground pipeline management with reliability and cost factors using Monte Carlo simulation. J. Loss Prevent. Process Industries. **63**, 104035 (2020)
68. S. Alessandri, A.C. Caputo, D. Corritore, R. Giannini, F. Paolacci, H.N. Phan, Probabilistic risk analysis of process plants under seismic loading based on Monte Carlo simulations. J. Loss Prev. Process Ind. **53**, 136–148 (2018)
69. R.T. Clemen, R.L. Winkler, Combining probability distributions from experts in risk analysis. Risk Anal. **19**(2), 187–203 (1999)
70. T. Rosqvist, *On the Use of Expert Judgement in the Qualification of Risk Assessment*, VTT Technical Research Centre of Finland (2003)
71. E. Zarei, B. Ramavandi, A.H. Darabi, M. Omidvar, A framework for resilience assessment in process systems using a fuzzy hybrid MCDM model. J. Loss Prevent. Process Industries **69**, 104375 (2021)
72. B. Ramavandi, A. H. Darabi, M. Omidvar, Risk assessment of hot and humid environments through an integrated fuzzy AHP-VIKOR method. Stochastic Environ. Res. Risk Assessment, 1–14 (2021)
73. B. Skelton, Process safety analysis: an introduction: IChemE (1997)

74. D. A. Crowl, J. F. Louvar, Chemical process safety: fundamentals with applications, Pearson Education (2001)
75. R. Ferdous, F. Khan, B. Veitch, P.R. Amyotte, Methodology for computer aided fuzzy fault tree analysis. Process Saf. Environ. Prot. **87**(4), 217–226 (2009)
76. K. Sentz, S. Ferson, Combination of evidence in Dempster-Shafer theory: Sandia National Laboratories Albuquerque (2002)
77. R. Ouache, R. Ruparathna, R. Sadiq, K. Hewage, Fire risk assessment model for residential buildings using bow-tie method. The University of British Columbia (2018)
78. D. Yuhua, Y. Datao, Estimation of failure probability of oil and gas transmission pipelines by fuzzy fault tree analysis. J. Loss Prev. Process Ind. **18**(2), 83–88 (2005)
79. H. Tanaka, L. Fan, F. Lai, K. Toguchi, Fault-tree analysis by fuzzy probability. IEEE Trans. Reliab. **32**(5), 453–457 (1983) .
80. K.B. Misra, G.G. Weber, Use of fuzzy set theory for level-I studies in probabilistic risk assessment. Fuzzy Sets Syst. **37**(2), 139–160 (1990)
81. D. Singer, A fuzzy set approach to fault tree and reliability analysis. Fuzzy Sets Syst. **34**(2), 145–155 (1990)
82. R. Kenarangui, Event-tree analysis by fuzzy probability. IEEE Trans. Reliab. **40**(1), 120–124 (1991)
83. J.P. Sawyer, S. Rao, Fault tree analysis of fuzzy mechanical systems. Microelectron. Reliab. **34**(4), 653–667 (1994)
84. P.V. Suresh, A.K. Babar, V.V. Raj, Uncertainty in fault tree analysis: A fuzzy approach. Fuzzy Sets Syst. **83**(2), 135–141 (1996)
85. S. Rivera, J. H. Baron, Using fuzzy arithmetic in containment event trees (2000)
86. E. Clifton (ed), Fault tree analysis-a history. Proceedings of the 17th International Systems Safety Conference (1999)
87. U. Hauptmanns, Fault tree analysis of a proposed ethylene vaporization unit. Ind. Eng. Chem. Fundam. **19**(3), 300–309 (1980)
88. U. Hauptmanns, *Fault tree analysis for process plants* (CRC Press, Engineering risk and hazard assessment, 2018), pp. 21–60
89. K. Hiromitsu, J. H. Ernest, Fault tree construction. probabilistic risk assessment and management for engineers and scientists, IEEE, pp 165–226 (1996)
90. D. Huang, T. Chen, M.J.J. Wang, A fuzzy set approach for event tree analysis. Fuzzy Sets Syst. **118**(1), 153–165 (2001)
91. C.T. Lin, M.J.J. Wang, Hybrid fault tree analysis using fuzzy sets. Reliab. Eng. Syst. Safety **58**(3), 205–213 (1997)
92. S. Kabir, Y. Papadopoulos, A review of applications of fuzzy sets to safety and reliability engineering. Int. J. Approximate Reasoning **100**, 29–55 (2018)
93. H. Furuta, N. Shiraishi, Fuzzy importance in fault tree analysis. Fuzzy Sets Syst. **12**(3), 205–213 (1984)
94. A.C.F. Guimarẽes, N.F.F. Ebecken, FuzzyFTA: a fuzzy fault tree system for uncertainty analysis. Ann. Nucl. Energy **26**(6), 523–532 (1999)
95. S.M. Lavasani, A. Zendegani, M. Celik, An extension to Fuzzy Fault Tree Analysis (FFTA) application in petrochemical process industry. Process Saf. Environ. Prot. **93**, 75–88 (2015)
96. J.H. Purba, D.T. Sony Tjahyani, A.S. Ekariansyah, H. Tjahjono, Fuzzy probability based fault tree analysis to propagate and quantify epistemic uncertainty. Ann. Nucl. Energy **85**, 1189–1199 (2015)
97. J.H. Purba, A fuzzy-based reliability approach to evaluate basic events of fault tree analysis for nuclear power plant probabilistic safety assessment. Ann. Nuclear Energy **70**(3), 21–29 (2014)
98. M. Abdelgawad, A.R. Fayek, Fuzzy reliability analyzer: quantitative assessment of risk events in the construction industry using fuzzy fault-tree analysis. J. Constr. Eng. Manag. **137**(4), 294–302 (2011)
99. A. Cheliyan, S. Bhattacharyya, Fuzzy fault tree analysis of oil and gas leakage in subsea production systems. J. Ocean Eng. Sci. **3**(1), 38–48 (2018)

100. Y.A. Mahmood, A. Ahmadi, A.K. Verma, A. Srividya, U. Kumar, Fuzzy fault tree analysis: a review of concept and application. Int. J. Syst. Assurance Eng. Manage. **4**(1), 19–32 (2013)
101. A. Ardeshir, M. Amiri, Y. Ghasemi, M. Errington, Risk assessment of construction projects for water conveyance tunnels using fuzzy fault tree analysis. Int. J. Civil Eng. **12**(4), 396–412 (2014)
102. X. Bian, C. Mou, Z. Yan, J. Xu (eds) Reliability analysis of AUV based on fuzzy fault tree. 2009 International conference on mechatronics and automation, IEEE, (2009)
103. Z. Chen, X. Wu, J. Qin, Risk assessment of an oxygen-enhanced combustor using a structural model based on the FMEA and fuzzy fault tree. J. Loss Prev. Process Ind. **32**, 349–357 (2014)
104. R. Flage, P. Baraldi, E. Zio, T. Aven, Probability and possibility-based representations of uncertainty in fault tree analysis. Risk Anal.: Int. J. **33**(1), 121–133 (2013)
105. M. Gierczak, The quantitative risk assessment of MINI, MIDI and MAXI horizontal directional drilling projects applying fuzzy fault tree analysis. Tunn. Undergr. Space Technol. **43**, 67–77 (2014)
106. K. B. Kim, H. J. Shim, M. S. Jae, *Fuzzy Uncertainty Evaluation for Fault Tree Analysis* (2015)
107. P.K.S. Singh, Fuzzy fault tree analysis using level (λ, ρ) interval-valued fuzzy numbers. Industrial Eng. Lett., **5**(2) (2015)
108. A. Mhalla, S. Collart Dutilleul, E. Craye, M. Benrejeb, Estimation of failure probability of milk manufacturing unit by fuzzy fault tree analysis. J. Intell. Fuzzy Syst. **26**(2), 741–750 (2014)
109. S. Rajakarunakaran, A.M. Kumar, V.A. Prabhu, Applications of fuzzy faulty tree analysis and expert elicitation for evaluation of risks in LPG refuelling station. J. Loss Prev. Process Ind. **33**, 109–123 (2015)
110. Y.E. Senol, B. Sahin, A novel real-time continuous fuzzy fault tree analysis (RC-FFTA) model for dynamic environment. Ocean Eng. **127**, 70–81 (2016)
111. M. Yazdi, E. Zarei, Uncertainty handling in the safety risk analysis: an integrated approach based on fuzzy fault tree analysis. J. Fail. Anal. Prev. **18**(2), 392–404 (2018)
112. Z.-J. Zhang, Z.-J. Wang, B.-H. Zhang (eds), Studies on median value of fuzzy numbers based on confidence level. 2008 International Conference on Machine Learning and Cybernetics, IEEE (2008)
113. S.A. Zonouz, S. G. Miremadi (eds), A fuzzy-Monte Carlo simulation approach for fault tree analysis. RAMS'06 Annual reliability and maintainability symposium, 2006, IEEE (2006)
114. H. Abdo, J.-M. Flaus, Monte Carlo simulation to solve fuzzy dynamic fault tree. IFAC-PapersOnLine. **49**(12), 1886–1891 (2016)
115. R. Duan, J. Fan, Reliability evaluation of data communication system based on dynamic fault tree under epistemic uncertainty. Math. Problems Eng., **2014** (2014)
116. R. Duan, H. Zhou, Diagnosis strategy for micro-computer controlled straight electro-pneumatic braking system using fuzzy set and dynamic fault tree. Eksploatacja i Niezawodność, **16**(2) (2014)
117. H. Huang, Y. Li, J. Sun, Y. Yang, N. Xiao, Fuzzy dynamic fault tree analysis for the solar array drive assembly. J. Mech. Eng. **49**(19), 70–76 (2013)
118. S. Kabir, E. Edifor, M. Walker, N. Gordon (eds), Quantification of temporal fault trees based on fuzzy set theory. Proceedings of the Ninth International Conference on Dependability and Complex Systems DepCoS-RELCOMEX June 30–July 4, 2014, (Springer, Brunów, Poland, 2014)
119. J. Tu, R. Cheng, Q. Tao, Reliability analysis method of safety-critical avionics system based on dynamic fault tree under fuzzy uncertainty. Eksploatacja i Niezawodność., **17**(1) (2015)
120. C. Yuyan, L. Ting, W. Jian, X. Rong, W. Xinmin, Fuzzy dynamic fault tree analysis for electromechanical actuator based on algebraic model with common-cause failures. Autom. Control. Comput. Sci. **50**(2), 80–90 (2016)
121. J.D. Andrews, S.J. Dunnett, Event-tree analysis using binary decision diagrams. IEEE Trans. Reliab. **49**(2), 230–238 (2000)
122. O. Nývlt, M. Rausand, Dependencies in event trees analyzed by Petri nets. Reliab. Eng. Syst. Saf. **104**, 45–57 (2012)

123. F. Lees, *Lees' Loss Prevention in the Process Industries: Hazard Identification, Assessment and Control* (Butterworth-Heinemann, 2012)
124. S. Mannan, Hazard assessment, in *Lees' Loss Prevention in the Process Industries*, 4th edn., ed. by S. Mannan (Butterworth-Heinemann, Oxford, 2012), pp. 284–404
125. R. Sadiq, E. Saint-Martin, Y. Kleiner, Predicting risk of water quality failures in distribution networks under uncertainties using fault-tree analysis. Urban Water J. 5(4), 287-304 (2008)
126. S. Ferson, J. Hajagos, D. Berleant, J. Zhang, W.T. Tucker, L. Ginzburg et al., Dependence in Dempster-Shafer theory and probability bounds analysis. Sandia Nat. Lab. (2004)
127. N.J. Duijm, Safety-barrier diagrams as a safety management tool. Reliab. Eng. Syst. Saf. **94**(2), 332–341 (2009)
128. Gareth (ed), Lessons learned from real world application of the bow-tie method. SPE Middle East Health, Safety, Security, and Environment Conference and Exhibition, OnePetro (2012)
129. W.E. Vesely, F. F. Goldberg, N.H. Roberts, D. F. Haasl, Fault tree handbook. (Nuclear Regulatory Commission Washington DC, 1981)
130. F.-R. Chevreau, J.-L. Wybo, D. Cauchois, Organizing learning processes on risks by using the bow-tie representation. J. Hazard. Mater. **130**(3), 276–283 (2006)
131. C. Delvosalle, C. Fievez, A. Pipart, B. Debray, ARAMIS project: A comprehensive methodology for the identification of reference accident scenarios in process industries. J. Hazard. Mater. **130**(3), 200–219 (2006)
132. J. Wilday, N. Paltrinieri, R. Farret, J. Hebrard, L. Breedveld, Addressing emerging risks using carbon capture and storage as an example. Process Saf. Environ. Prot. **89**(6), 463–471 (2011)
133. A. Tugnoli, Z. Gyenes, L. Van Wijk, M. Christou, G. Spadoni, V. Cozzani, Reference criteria for the identification of accident scenarios in the framework of land use planning. J. Loss Prev. Process Ind. **26**(4), 614–627 (2013)
134. T. Deacon, P. Amyotte, F. Khan, S. MacKinnon, A framework for human error analysis of offshore evacuations. Saf. Sci. **51**(1), 319–327 (2013)
135. S. Rathnayaka, F. Khan, P. Amyotte, Risk-based process plant design considering inherent safety. Saf. Sci. **70**, 438–464 (2014)
136. L. Lu, W. Liang, L. Zhang, H. Zhang, Z. Lu, J. Shan, A comprehensive risk evaluation method for natural gas pipelines by combining a risk matrix with a bow-tie model. J. Natural Gas Sci. Eng. **25**, 124–133 (2015)
137. D. Siuta, A. Kotynia, B. Kukfisz, W. Szaferski (eds), Risk assessment in process industries using fuzzy logic approach. *Seminar on Practical Aspects of Chemical Engineering* (Springer, Cham, 2019)
138. M. Abimbola, F. Khan, N. Khakzad, Dynamic safety risk analysis of offshore drilling. J. Loss Prev. Process Ind. **30**, 74–85 (2014)
139. A. Badreddine, T.B. Romdhane, M.A.B. HajKacem, N.B. Amor, A new multi-objectives approach to implement preventive and protective barriers in bow tie diagram. J. Loss Prev. Process Ind. **32**, 238–253 (2014)
140. A. Badreddine, N.B. Amor, A Bayesian approach to construct bow tie diagrams for risk evaluation. Process Saf. Environ. Prot. **91**(3), 159–171 (2013)
141. A. Badreddine, N. B. Amor (eds), A new approach to construct optimal bow tie diagrams for risk analysis. *International Conference on Industrial, Engineering and Other Applications of Applied Intelligent Systems* (Springer, 2010)
142. C. Delvosalle, C. Fiévez, A. Pipart, J.C. Fabrega, E. Planas, M. Christou et al., Identification of reference accident scenarios in SEVESO establishments. Reliab. Eng. Syst. Saf. **90**(2–3), 238–246 (2005)
143. A. Franks, R. Whitehead, P. Crosthwaite, L. Smail, Application of QRA in operational safety issues, HSE Books (2002)
144. IADC, *Health Safety and Environment Case Guidelines for Mobile Offshore Drilling Units.* (Houston, International Association of Drilling Contractors, 2010)
145. N. Khakzad, F. Khan, P. Amyotte, Dynamic safety analysis of process systems by mapping bow-tie into Bayesian network. Process Saf. Environ. Prot. **91**(1–2), 46–53 (2013)

146. N. Khakzad, F. Khan, P. Amyotte, Dynamic risk analysis using bow-tie approach. Reliab. Eng. Syst. Saf. **104**, 36–44 (2012)
147. N. Khakzad, F. Khan, P. Amyotte, Quantitative risk analysis of offshore drilling operations: a Bayesian approach. Saf. Sci. **57**, 108–117 (2013)
148. N. Paltrinieri, F. Khan, Dynamic risk analysis in the chemical and petroleum industry: evolution and interaction with parallel disciplines in the perspective of industrial application, (Butterworth-Heinemann, 2016)
149. A. Targoutzidis, Incorporating human factors into a simplified "bow-tie" approach for workplace risk assessment. Saf. Sci. **48**(2), 145–156 (2010)
150. K. Mokhtari, J. Ren, C. Roberts, J. Wang, Application of a generic bow-tie based risk analysis framework on risk management of sea ports and offshore terminals. J. Hazard. Mater. **192**(2), 465–475 (2011)
151. E. Zarei, F. Khan, M. Yazdi, A dynamic risk model to analyze hydrogen infrastructure. Int. J. Hydrogen Energy **46**(5), 4626–4643 (2021)
152. L.A. Zadeh, Fuzzy sets. Inf. Control **8**(3), 338–353 (1965)
153. L.A. Zadeh, The concept of a linguistic variable and its application to approximate reasoning—I. Inf. Sci. **8**(3), 199–249 (1975)
154. L.A. Zadeh, The concept of a linguistic variable and its application to approximate reasoning-III. Inf. Sci. **9**(1), 43–80 (1975)
155. L.A. Zadeh, Toward a theory of fuzzy information granulation and its centrality in human reasoning and fuzzy logic. Fuzzy Sets Syst. **90**(2), 111–127 (1997)
156. M.S. Anggreainy, M.R. Widyanto, B.H. Widjaja, N. Soedarsono, Gaussian fuzzy number for STR-DNA similarity calculation involving familial and tribal relationships. Adv. Bioinform. **2018**, 8602513 (2018)
157. L.A. Zadeh, The concept of a linguistic variable and its application to approximate reasoning—II. Inf. Sci. **8**(4), 301–357 (1975)
158. G. Zheng, N. Zhu, Z. Tian, Y. Chen, B. Sun, Application of a trapezoidal fuzzy AHP method for work safety evaluation and early warning rating of hot and humid environments. Saf. Sci. **50**(2), 228–239 (2012)
159. L.A. Zadeh, *A Computational Approach to Fuzzy Quantifiers in Natural Languages* (Elsevier, Computational linguistics, 1983), pp. 149–184
160. W. Karwowski, A. Mital, Potential applications of fuzzy sets in industrial safety engineering. Fuzzy Sets Syst. **19**(2), 105–120 (1986)
161. J.B. Bowles, C.E. Pelaez, Application of fuzzy logic to reliability engineering. Proc. IEEE **83**(3), 435–449 (1995)
162. D.J. Dubois, Fuzzy sets and systems: theory and applications, (Academic press, 1980)
163. D. Dubois, Possibility theory and statistical reasoning. Comput. Stat. Data Anal. **51**(1), 47–69 (2006)
164. M. Zahabi, D. Kaber, A fuzzy system hazard analysis approach for human-in-the-loop systems. Saf. Sci. **120**, 922–931 (2019)
165. S. Kambalimath, P.C. Deka, A basic review of fuzzy logic applications in hydrology and water resources. Appl. Water Sci. **10**(8), 191 (2020)
166. E.H. Mamdani, S. Assilian, An experiment in linguistic synthesis with a fuzzy logic controller. Int. J. Man Mach. Stud. **7**(1), 1–13 (1975)
167. P.M. Larsen, Industrial applications of fuzzy logic control. Int. J. Man Mach. Stud. **12**(1), 3–10 (1980)
168. T. Takagi, M. Sugeno, Fuzzy identification of systems and its applications to modeling and control. IEEE Trans. Syst. Man Cybern. **1**, 116–132 (1985)
169. H.T. Liu, Y.I. Tsai, A fuzzy risk assessment approach for occupational hazards in the construction industry. Safety Sci. **50**(4), 1067–1078 (2012)
170. B. Ruge (ed), Risk matrix as tool for risk assessment in the chemical process industries. Probabilistic Safety Assessment and Management, (Springer, 2004)
171. M. Detyniecki, D. B. Bouchon-meunier, D. R. Yager, R. H. Prade, Mathematical aggregation operators and their application to video querying (2000)

172. R.R. Yager, On ordered weighted averaging aggregation operators in multicriteria decision-making. IEEE Trans. Syst. Man Cybern. **18**(1), 183–190 (1988)
173. T. Tsabadze, A method for fuzzy aggregation based on group expert evaluations. Fuzzy Sets Syst. **157**(10), 1346–1361 (2006)
174. B. Schweizer, A. Sklar. Probabilistic metric spaces. Courier Corporation (2011)
175. M. Sugeno, Theory of fuzzy integrals and its applications. Doct Thesis, Tokyo Institute of technology (1974)
176. W. Silvert, Symmetric summation: a class of operations on fuzzy sets. Readings in Fuzzy Sets for Intelligent Systems. (Elsevier, 1993), pp. 77–9
177. H.-S. Lee, Optimal consensus of fuzzy opinions under group decision making environment. Fuzzy Sets Syst. **132**(3), 303–315 (2002)
178. R. R. Yager, A new methodology for ordinal multiobjective decisions based on fuzzy sets. Readings in Fuzzy Sets for Intelligent Systems. (Elsevier, 1993), pp 751–6
179. C. Lu, J. Lan, Z. Wang, Aggregation of fuzzy opinions under group decision-making based on similarity and distance. J. Syst. Sci. Complexity **19**(1), 63–71 (2006)
180. C. Kim, Y. Ju, M. Gens, Multilevel fault tree analysis using fuzzy numbers. Comput. Oper. Res. **23**(7), 695–703 (1996)
181. L.A. Zadeh, Fuzzy sets as a basis for a theory of possibility. Fuzzy Sets Syst. **1**(1), 3–28 (1978)
182. H. Langseth, L. Portinale, Bayesian networks in reliability. Reliab. Eng. Syst. Saf. **92**(1), 92–108 (2007)
183. V. Renjith, G. Madhu, V.L.G. Nayagam, A. Bhasi, Two-dimensional fuzzy fault tree analysis for chlorine release from a chlor-alkali industry using expert elicitation. J. Hazard. Mater. **183**(1–3), 103–110 (2010)
184. T. Onisawa, An approach to human reliability in man-machine systems using error possibility. Fuzzy Sets Syst. **27**(2), 87–103 (1988)
185. T. J. Ross, Properties of membership functions, fuzzification, and defuzzification. Fuzzy Logic Eng. Appl., 89–116 (2010)
186. M. Omidvar, A. Mazloumi, I. Mohammad Fam, F. Nirumand, Development of a framework for resilience measurement: suggestion of fuzzy Resilience Grade (RG) and fuzzy Resilience Early Warning Grade (REWG). Work **56**(3), 463–474 (2017)
187. CCPS, *Layer of Protection Analysis: Simplified Process Risk Assessment* (2011)
188. A. Nassaj, J. Barabady (eds). Fault tree analysis of oil and gas distillation tower and application of Bayesian Networks. 2016 IEEE international conference on industrial engineering and engineering management (IEEM), IEEE (2016)
189. J.A. Vílchez, V. Espejo, J. Casal, Generic event trees and probabilities for the release of different types of hazardous materials. J. Loss Prev. Process Ind. **24**(3), 281–287 (2011)

Chapter 4
Optimizing the Allocation of Risk Control Measures Using Fuzzy MCDM Approach: Review and Application

Mostafa Pouyakian, Ashkan Khatabakhsh, Mohammad Yazdi, and Esmaeil Zarei

4.1 Introduction

Choosing the best decision has been one of the oldest intellectual challenges of mankind. In today's modern and complex world, decision making has become a specialized and difficult subject. Managers and scholars are often surrounded by large amounts of data and are pressured by interests or conflicting orientations of decision criteria. They must be able to make the best decisions and defend the credibility of their choices. This chapter addresses optimizing the selection of risk control measures.

4.2 Typical Framework for Risk Management

Risk management is a process that analyzes, forecasts and controls risk to reduce losses. This process consists of three basic steps: hazard identification, risk assessment, and risk control. Over the past five decades, many quantitative, qualitative, and hybrid methods have been developed and referenced for the first two steps [1–7]. HAZOP (Hazard and Operability Study), FMEA (Failure Modes and Effects Analysis), ETBA (Energy Trace and Barrier Analysis), What if analysis, and PHA

M. Pouyakian (✉) · A. Khatabakhsh
Department of Occupational Health and Safety Engineering, School of Public Health and Safety, Shahid Beheshti University of Medical Sciences, Tehran, Iran

M. Yazdi
Faculty of Engineering and Applied Science, Memorial University of Newfoundland, St. John's, NL A1B 3X5, Canada

E. Zarei
Centre for Risk, Integrity and Safety Engineering (C-RISE), Faculty of Engineering and Applied Science, Memorial University of Newfoundland, St. John's, NL A1B 3X5, Canada

M. Yazdi, *Linguistic Methods Under Fuzzy Information in System Safety and Reliability Analysis*, Studies in Fuzziness and Soft Computing 414,
https://doi.org/10.1007/978-3-030-93352-4_4

Table 4.1 A typical worksheet to record the risk analysis data

No	Group 1: Hazard data (source, type, activation way, intermediate and systemic effect)	Group 2: Risk data before intervention (Probability, consequences, detection rate, exposure)	Group 3: Risk control data (recommended counter measures)	Group 4: Risk data after intervention (Probability, consequences, detection rate, exposure)
1				
2				
...				
n				

(Preliminary Hazard Analysis) are some of the hazard identification techniques that have been frequently applied by safety professionals, especially in the process industries. Also, some techniques have been developed to evaluate the risk based on two main elements: probability and consequences. While some techniques do this based on an extra element like exposure or detection rate. MIL-STD-882 is a well-known example of the two-element risk assessment method and Risk Priority Number (RPN) is a famous one for a three-element risk assessment.

Generally, assessors record the results of an assessment in a specified worksheet. Table 4.1 shows a typical risk analysis worksheet. Four groups of data are recorded in the worksheets. (1) Hazard data including type and source of hazard, activation way, the intermediate and systemic effect of hazard are recorded in the first group of columns of the worksheet in terms of the requirements of the selected method. (2) The risk data includes probability, consequences, exposure, and detectability, which are recorded in the second group of columns based on the requirements of the selected risk assessment method. (3) When the risks are beyond the acceptable limits of the system, the assessment team should offer some risk control measures to mitigate the risk level below the acceptable level. Hence, the last column or one of the last columns of the worksheet is called the "Risk Control Measures (RCMs)", "recommended action", "mitigation method", "mitigation recommendation", or simply "recommendations" [8]. In this column, some technical, administrative, or personal solutions are recommended to reduce the estimated risk below the acceptable levels. (4) Finally, the assessment team must assess the impact of the proposed measures to reduce the level of risk to an acceptable level.

4.3 Risk Control Models

The three most popular and common safety control models are "hierarchy of controls" [9, 10], "chemical process safety strategies" [11], and Haddon matrix [12, 13]. These models organize the way safety professionals and decision makers think about risk mitigation methods.

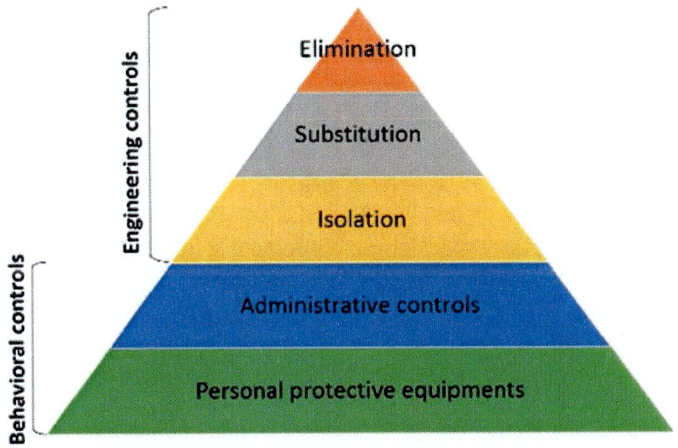

Fig. 4.1 Hierarchy of controls

4.3.1 Hierarchy of Control Model

In the hierarchy of the control model (Fig. 4.1), control measures fit into five levels of descending effectiveness, which are Elimination, Substitution, Isolation, Administration, and Personal protective equipment (PPE). Conversely, the applicability of control levels decreases from end to beginning. This means that using personal protective equipment against hazards is much easier than eliminating hazards. The top three levels in the control triangle are intended to control the hazards at the workplace based on technological interventions while the latter two levels are behavioral controls that limit workers' exposure to hazards [14].

4.3.2 Chemical Process Safety Strategies

The control measures in "Chemical Process Safety Strategies" are grouped into four categories in order of their robustness: inherently safer design (ISD), passive controls, active controls, and procedural controls (Fig. 4.2). The first three can be characterized

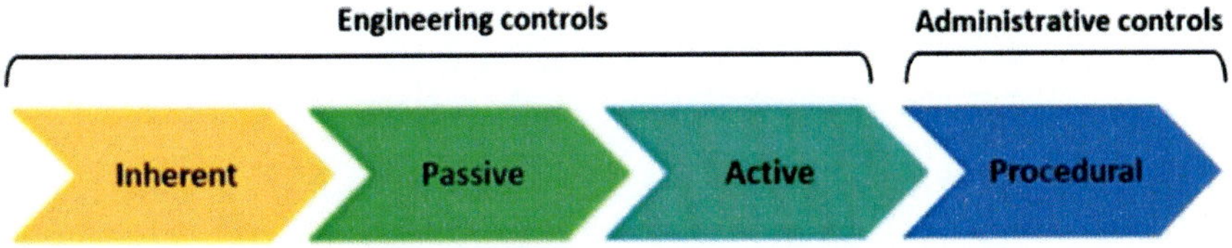

Fig. 4.2 Chemical process safety strategies

as engineering controls, while the last (procedural) can be characterized as administrative control [11]. ISD was developed 40 years ago in the processing industries after the publication of the Flixborough accident report and aimed to reduce the probability and severity of risks while also eliminating hazards by making major changes to the design and planning of the systems process (i.e., using less or no hazardous material or processes). ISD eliminates or reduces risks by substitution, minimizing, moderating, and simplifying methods [15]. The passive strategy consists of engineering controls that reduce risks by designing equipment without an active device. The active strategy controls risk by implementing active devices (sensors, sprinklers…). The procedural strategy implements administrative controls such as instructions, standards, and training [11, 16]. ISD and passive strategy have high reliability and priority due to their reliance on non-active devices and their effect on processes and equipment design. The concept of "Chemical Process Safety Strategies" have been used in Layers of Protection Analysis method (LOPA) as "Independent Protection Layers" [17] (Fig. 4.3).

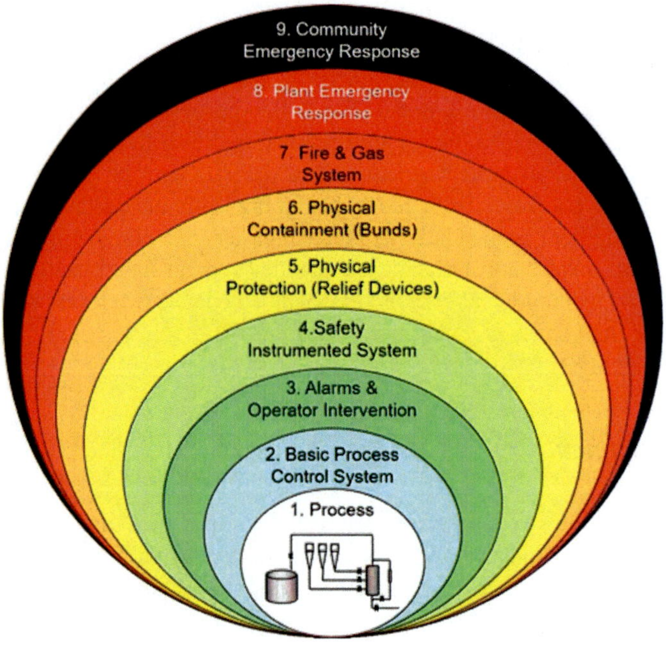

Fig. 4.3 LOPA onion of control

4.3.3 Haddon's Matrix

William Haddon Jr. developed a time and target-based model in the 1970s to categorize the safety countermeasures to control the accidents in the road traffic domain [12]. He introduced a 12-cell matrix including four columns (human, machine/equipment, physical environment, social environment) and three rows (Pre-event, Event, Post-event) (Fig. 4.4). The column names may slightly vary in different studies. Each cell represents the time phase and valuable target that the control measures impact. For example, human factor goes back to human-related actions like training on work safety, how to deal with accidents, first aid measures, etc.

Machine/equipment factors include RCMs that have a control effect on machines and equipment such as improving the design, dead man's switches, and fuses. Physical environment refers to fireproof coatings and emergency exit routes, etc. Socio-economic factors refer to rules, guidelines, budgets, etc. Rows divide the RCMs according to phase time (i.e., the time that RCM applies its corrective effect) to pre-event, during event, and post-event groups. The goal in the pre-event phase is to control factors that lead to incidents. The goal in the event phase is identifying, limiting, blocking and/or modifying the hazardous processes that have already been initiated or have been predicted. The goal in the post-event phase is repair, reconstruction, and rehabilitation post-hazardous processes [18, 19]. By identifying interventions that fit within each cell of the matrix one can generate a list of strategies to address a variety of injuries or damages. This concept has been used in other domains like the healthcare and chemicals industry [13].

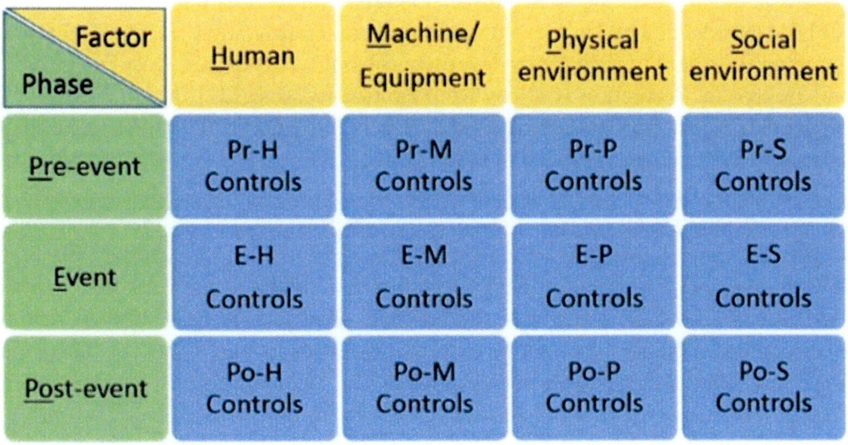

Fig. 4.4 Haddon matrix

4.4 Evaluation of Risk Control Measures: Conflicting Criteria

Safety professionals gather hazard data and risk data through systematic, complicated, and sophisticated methods [20, 21]. A couple of advanced methods recently have been presented to dynamic modeling of the probability of events and consequence analysis. Unlike the first two steps of the risk management process, there are few methods developed for the third step (risk control). Most of the time, suggested corrective actions are generally based on the dominant thinking of the hierarchy of controls as well as the level of knowledge and experience of the members of the assessment team. Using risk control models helps engineers and managers to regulate their thoughts in the choice of risk control measures. Unfortunately, in their current form, these models do not provide a systematic way to evaluate the RCMs.

RCMs must be able to control at least one of the key elements of risk, including consequences, probability of occurrence, exposure to the hazard, or detection rate. However, in most of the risk assessment techniques, RCMs are not being evaluated and categorized based on a theoretical method [22, 23]. In other words, in a very few classic risk assessment techniques, a systematic method has been provided to evaluate the RCMs [24]. This problem is continuing in recently developed safety analysis methods. Zarei et al., developed a hybrid model for human factor analysis in process accidents (natural gas pipeline) by using HFACS, FST and expert elicitation, Fuzzy AHP, and BN. They identified key safety measures to improve safety based on preventing reoccurrence of similar accidents, but there were no criteria for prioritizing safety measures. [25]

One of the first attempts to make decisions about the RCMs was made by William [26]. He introduced J-index to evaluate the economic justification of control measures based on risk score (derived from severity, probability, exposure), Cost Factor (CF), and Degree of Correction (DC) by each control measure [27] in which a J value higher than or equal to 10 is considered an acceptable RCM and a value smaller than 10 is considered as an unacceptable RCM.

In a recent study on assessing the sufficiency of control measures, Zhao et al., studied the quality of control measures for electrical hazards in the construction sector. The objective of their study was to analyze existing control measures of electrical hazards by applying an approach of the hierarchy of controls. They created "tag clouds" based on a big data source of used control measures for preventing electrocution in the US construction industry from 1989 to 2012. They concluded that behavioral controls (two lower levels of the hierarchy of controls) are still dominant control measures regardless of technological progress over the years. They used a simple weighting method to rate the five levels of the hierarchy of controls (5 for most effective control measures i.e., elimination of hazard, and 1 for least effective control measures i.e., personal protective equipment). Their rating method was based on just the "effectiveness" criterion [14].

Some constraints, like the available budget for a project, make it necessary to assess the cost-effectiveness of any implemented countermeasures [28]. Atehnjia et al., used Bayesian networks for the cost–benefit analysis of RCMs in docking operations. Their method evaluates RCMs through Bayesian networks and verbal expressions of cost, benefit, risk reduction, reference values, and uncertainty criteria [29].

A researcher suggested a hazard intervention effectiveness matrix that consisted of hazard control hierarchy principles and Haddon's theory [30]. In addition, each RCM was qualitatively evaluated using the "implemented as intended", "relatively stable" and "seem to be achieving positive results" criteria. Runyan introduced a three-dimension matrix to facilitate the use of Haddon's matrix in the decision-making process and to help choose the appropriate control measures [13]. The third dimension in her model was "decision criteria", but she did not investigate the importance of suggested criteria for choosing an RCM. Also, the relative importance of Haddon's Matrix components to each other has not been addressed in previous studies [31, 32].

While "effectiveness" is the main criterion for any decision on control measures [13], there are some other criteria that affect the overall result. Selecting a control measure among a couple of suggested measures is influenced by various factors such as the cost of implementing the measure, the implementing time phase, required time for implementing the measure, reliability, usability, equipment (facilities), and so on [26]. Yazdi et al., developed a probabilistic safety analysis model in process systems-based DEMATEL (decision making trial and evaluation laboratory) approach. The model was used for prioritizing critical root events and corrective actions. Criteria for prioritizing corrective actions were cost, benefit, and efficiency [33, 34].

Runyan (1998) suggested eight criteria for prioritizing control measures including effectiveness, cost, freedom, equity, stigmatization, preferences of the affected community or individuals, and feasibility. She noticed that these criteria are not the final ones and can be varied according to the individual problem settings [13] (Fig. 4.5).

Problems like applicability as well as technical compatibility and related standards are important limitations that affect decisions on RCMs. Since various criteria affect the process of selecting optimized risk control measures, choosing an appropriate set of control measures to reduce risks is challenging work. In many cases, there are serious doubts about the preference of control measures over each other. Decision making under conditions of high uncertainty results in unpredictability. Hence, managers and safety professionals encounter a complicated optimizing problem.

4.5 Multi-Criteria Decision-Making Methods

An optimization problem is a difficulty of finding the best solution from all feasible solutions. Multi-Criteria Decision-making Methods (MCDM) is a generic term for a group of techniques that find the optimal answer where there are difficulties in ranking alternatives according to multiple complex and conflicting criteria [35]. Since the 1960s, MCDM has grown as a part of science, that deals with designing computational and mathematical tools for choosing the best alternatives by decision-makers

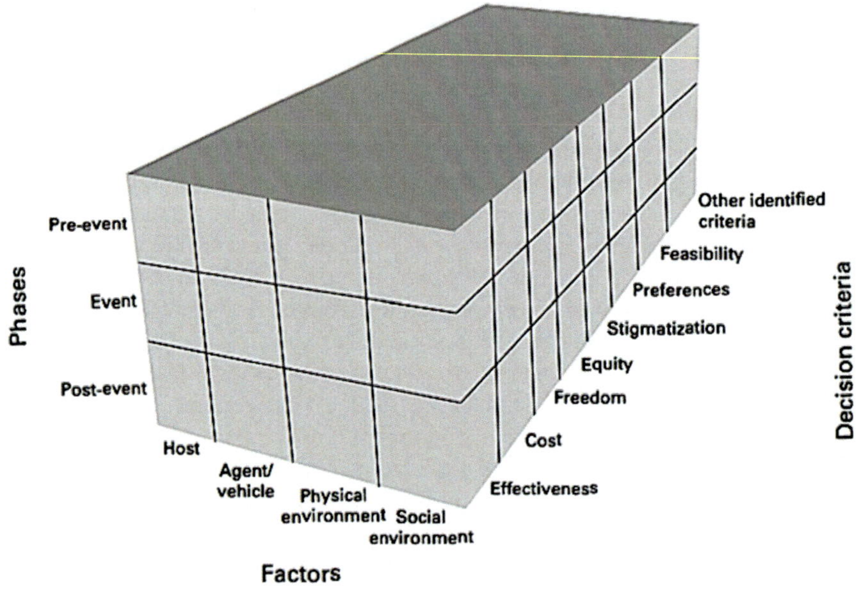

Fig. 4.5 Runyan's proposed three dimensional Haddon Matrix [13]

when there is a complex situation [36, 37]. After the 1990s, MCDM techniques grew at a fast rate [38].

MCDM techniques excessively use human participation and judgments [39]. They are characterized by alternatives, criteria against evaluated alternatives, scores of alternatives on the criteria, and criteria weights demonstrating the relative importance of each criterion in contrast with others [40]. In fact, the key aspect of MCDM techniques is their flexibility on the judgment of the decision-maker(s). Where many criteria have come into existence, the best one can be obtained by analyzing different scopes of the criteria, weights of the criteria, and the selection of the optimum ones using any MCDM techniques [41]. They survey the best decision by assigning performance scores and weights [42, 43].

MCDM methods can be broadly classified into two categories: discrete MCDM or discrete Multi-attribute Decision-Making (MADM) and continuous Multi-objective Decision-Making (MODM) methods [44]. Various MCDM techniques have been developed over the years; examples include the following: Decision Making Trial and Evaluation Laboratory (DEMATEL) [45], The Analytic Hierarchy Process (AHP) [46], Analytic Network Process (ANP) [47], Simple Additive Weighting (SAW), Elimination and Choice Expressing Reality (ELECTRE), the Linear Programming Technique for Multidimensional Analysis of Preference (LINMAP) [41], Best–Worst Method (BWM) [48], Preference Ranking Organization Method for Enrichment of Evaluations (PROMETHEE) [37], TODIM (an acronym in Portuguese for interactive multi-criteria decision making), the Technique for Order of Preference by Similarity to Ideal Solution (TOPSIS) [49], the VIKOR (an acronym in Serbian language for multicriteria optimization and compromise solution) [50].

4.6 Expert Opinions; Vague Linguistic Values

There are many situations in safety science wherein assessing a situation or making a decision for safe management of a process requires an expert opinion. These are situations about which there is usually no precise or quantitative data, or accessing quantitative data requires a lot of time and money. Therefore, experts generally express their opinions in linguistic non-numerous values such as high, low, medium, mild, severe, catastrophic, and so on. Also, they use adjectives and adverbs like "very" and "rather" to create additional linguistic values to refer to intermediate values [40].

The most well-known examples of such situations are estimating the probability of detecting a hazard or failure, estimating the probability of an accident, and assessing the severity of the possible consequences. Experts' opinions on the appropriateness of a solution are expressed in words such as appropriate, efficient, effective, weak, and so on. The ambiguity and uncertainty in these words make it difficult to decide which solution to choose, especially when two evaluators have different opinions.

Fuzzy sets theory offers a solution to overcome the vagueness of the words. This theory can convert much imprecise linguistic information into numerical data and provide the basis for reasoning, conclusion, and decision making in conditions of uncertainty. Unlike the Boolean logic that variables take only the integer values of zero and one, in fuzzy logic, the truth value of variables may be any real number between zero and one [51]. Thus, fuzzy mathematic is a solution to formulate the verbal data to determine which side the opinions are strongest.

The subjectivity in the decision-making process often leads to uncertainty due to the lack of information [43]. As a result, most MDCM methods have been developed with fuzzy sets theory. As a result, it could be possible to decrease the uncertainty environment [38, 52, 53].

4.7 Fuzzy Analytic Hierarchy Process (FAHP)

The Analytic Hierarchy Process (AHP) introduced by Saaty is one of the most common MCDM techniques and has been used widely in various fields [54]. This method is employed whenever several criteria for quantitative/qualitative decision-making exist or when there is no specific unit of measurement available [55]. The advantages of this method include sensitivity analysis, determining decisions' consistency ratio, strong theoretical basis [56], simplicity, ease of use, and high flexibility in various fields such as risk assessment and management [57]. The AHP theory tries to measure the relative importance of alternatives concerning each criterion by using pairwise comparisons. To do pairwise comparisons, experts use a conversion scale with crisp values to express their ideas. An important limitation of AHP is in using crisp values to reflect human thinking. Dealing with incomplete information, predictions, and vagueness in human thinking makes it possible to use fuzzy sets in the AHP method [58] hence, the FAHP method was developed to specify the subjective thinking style.

Van Laarhoven and Pedrycz were the first to apply fuzzy logic using the AHP method. They used a fuzzy logarithmic least square method to obtain triangular fuzzy weights [59]. After Van Laarhoven and Pedrycz various researchers introduced different FAHP methods [60–64]. In the present application, the Buckley method was used to obtain fuzzy weights from fuzzy pairwise comparisons since the calculations of the aforementioned methods are very complex or have limitations. For example, according to Kubler et al., the extent analysis method developed by Change despite its reputation does not use all the information from the triangular comparison matrix that could result in an irrational zero or negative weight to criteria or alternatives [39, 61].

The procedure of Buckley FAHP method is explained below [65]:

Step 1: Fuzzy pairwise comparison matrices are constructed for all the criteria and alternatives by using the triangular fuzzy conversion scale in Table 4.1. This matrix represents the results of fuzzy pairwise comparisons among criteria or alternatives concerning the upper agent according to the opinion of an expert. The matrix dimension is equal to the number of criteria or alternatives. Each matrix component indicates the priority degree of the matrix row to its respective column.

$$\tilde{A} = \begin{bmatrix} 1 & \tilde{a}_{12} & \cdots & \tilde{a}_{1n} \\ \tilde{a}_{21} & 1 & \cdots & \tilde{a}_{2n} \\ \vdots & \vdots & \ddots & \vdots \\ \tilde{a}_{n1} & \tilde{a}_{n2} & \cdots & 1 \end{bmatrix} \tag{4.1}$$

$$\tilde{a}_{ij} = \begin{cases} \tilde{1}, \tilde{2}, \tilde{3}, \tilde{4}, \tilde{5}, \tilde{6}, \tilde{7}, \tilde{8}, \tilde{9} & if\, C/A\, i\, is\, relatively\, important\, to\, C/A\, j \\ 1 & if\, i = j \\ \frac{1}{\tilde{1}}, \frac{1}{\tilde{2}}, \frac{1}{\tilde{3}}, \frac{1}{\tilde{4}}, \frac{1}{\tilde{5}}, \frac{1}{\tilde{6}}, \frac{1}{\tilde{7}}, \frac{1}{\tilde{8}}, \frac{1}{\tilde{9}} & if\, C/A\, i\, is\, less\, important\, to\, C/A\, j \end{cases} \tag{4.2}$$

Step 2: Since the opinions of various experts were used, multiple matrices were acquired. The Buckley fuzzy geometric mean method was used to consider the whole range of opinions in the group matrix. In the following formula, the superscripts refer to different experts and E represents the total number of experts:

$$\tilde{a}_{ij} = \left(\tilde{a}_{ij}^{1} \times \tilde{a}_{ij}^{2} \times \ldots \times \tilde{a}_{ij}^{E}\right)^{\frac{1}{E}} \tag{4.3}$$

Step 3: The fuzzy geometric mean values of the group matrix is defined by using the geometric mean method where n refers to the dimension of the matrix:

$$\tilde{r}_i = (\tilde{a}_{i1} \times \tilde{a}_{i2} \times \ldots \times \tilde{a}_{in})^{\frac{1}{n}} \tag{4.4}$$

Step 4: Fuzzy weights $(\tilde{w}_i)$ of each criterion or alternative is calculated. In the below criterion $\tilde{w}_i = (\mathrm{Lwi}, \mathrm{Mwi}, \mathrm{Uwi})$, where Lwi, Mwi Uwi represent the lower, middle, and upper values of the fuzzy triangular weight of criterion/alternative i.

$$\tilde{w}_i = \tilde{r}_i \times (\tilde{r}_1 + \tilde{r}_2 + \ldots + \tilde{r}_n)^{-1} \tag{4.5}$$

Step 5: The crisp (defuzzified) weight values of criterion/alternative i (w_i) is calculated by the center of area (COA) method as follows:

$$w_i = [((\mathrm{Uwi} - \mathrm{Lwi}) + (\mathrm{Mwi} - \mathrm{Lwi}))/3] + \mathrm{Lwi} \tag{4.6}$$

4.7.1 Consistency Ratio

Next, the consistency ratio (C.R) must be calculated for group matrices. C.R is a verification of the rational judgment performed by the experts. Due to the use of triangular fuzzy numbers, the consistency ratio could not be calculated by the method proposed by Saaty according to crisp numbers. To circumvent this problem, the Gogus and Boucher method was used to calculate the C.R of fuzzy matrices [66]. The procedure is as follows:

Step 1: Two matrices are constructed, one from the middle values of the fuzzy group and another from the geometric mean of lower and upper values of the fuzzy group.

Step 2: After acquiring matrices with crisp values, we then calculate the C.R according to the Saaty (1980) method. For this purpose, the matrix must be normalized and then the weight vector is calculated for the middle matrix $(\boldsymbol{w}_i^m)$ and geometric mean matrix $(\boldsymbol{w}_i^g)$ as follows:

$$\mathrm{W}^{\mathrm{m}} = [\boldsymbol{w}_i^m]\text{where } \boldsymbol{w}_i^m = \frac{1}{n}\sum_{j=1}^{n}\frac{a_{ijm}}{\sum_{i=1}^{n} a_{ijm}} \tag{4.7}$$

$$\mathrm{W}^{\mathrm{g}} = [\boldsymbol{w}_i^g]\text{where } \boldsymbol{w}_i^g = \frac{1}{n}\sum_{j=1}^{n}\frac{\sqrt{a_{iju}\cdot a_{ijl}}}{\sum_{i=1}^{n}\sqrt{a_{iju}\cdot a_{ijl}}} \tag{4.8}$$

Step 3: The largest eigenvalue for the middle matrix $\left(\lambda_{max}^{m}\right)$ and geometric mean matrix $\left(\lambda_{max}^{g}\right)$ are calculated as follows:

$$\lambda_{max}^{m} = \frac{1}{n}\sum_{i=1}^{n}\sum_{j=1}^{n} a_{ijm}\left(w_j^m / w_i^m\right) \tag{4.9}$$

$$\lambda_{max}^{g} = \frac{1}{n}\sum_{i=1}^{n}\sum_{j=1}^{n} \sqrt{a_{iju} \cdot a_{ijl}}\left(\frac{w_j^g}{w_i^g}\right) \tag{4.10}$$

Step 4: The consistency index that represents deviation from perfect consistency is calculated for the middle matrix (CI^m) and the geometric mean matrix (CI^g) as follows:

$$CI^m = \frac{\left(\lambda_{max}^{m} - n\right)}{(n-1)} \tag{4.11}$$

$$CI^g = \frac{\left(\lambda_{max}^{g} - n\right)}{(n-1)} \tag{4.12}$$

Step 5: The consistency ratio is calculated by dividing the consistency index by a random index (R.I). Since decimals are used in matrices and not crisp values in a conversion scale, we use R.I as suggested by Gogus and Boucher. According to Gogus and Boucher, for an acceptable fuzzy paired comparison matrix, the consistency ratios for both gained matrices should be less than 0.1 and if not, the comparison matrix must be revised, and the consistency ratio recalculated. If only one of the CRs is greater than 0.1 then the expert must re-evaluate the mean values while keeping extreme values fixed.

4.7.2 Absolute AHP

The absolute AHP developed by Saaty is a hierarchy consisting of goal, criteria, rating scale, and alternatives [67], while the common AHP (relative AHP) hierarchy consists of a goal, criteria, and alternatives. In the hierarchy of absolute AHP, the alternatives (fourth level) do not appear because the alternatives (RCMs) are not being pair compared concerning criteria but rated instead on the rating scale. According to Liberatore et al., this approach is usually used when there is a large number of alternatives (more than seven), or the number of alternatives is constantly increasing or decreasing. The use of the AHP method in these situations renders computational effort and pairwise comparison infeasible and uses a lot of time and resources. The method of assigning a rating scale to each criterion and evaluating each alternative independently of others makes absolute rating possible and eliminates the need for alternatives pairwise comparisons [68].

4.8 Application of MCDM Techniques for Safety Optimization Problems

MCDM techniques have been applied to many domains such as energy, environment, and sustainability, supply chain management, quality management, safety, and risk management [41]. Over the last decade, lots of attention has been paid to the application of MCDM in different parts of risk management. In this regard, Gul presents a review article on occupational health and safety risk assessment approaches based on multi-criteria decision-making techniques [69]. Khatabakhsh et al., used the ANP method to rank the best hazard identification techniques in process industries [70]. Omidvari et al., used Fuzzy Multiplicative Best–Worst Method (IFMBWM) in combination with FMEA method to assess the risk of fire in hospitals and health care centers [71]. Kaleh et al., provide an algorithm for ranking the safety of buildings to implement a tool to predict the safety status of institutional buildings in operation using the fuzzy Analytic Hierarchy Process (FAHP) method. Respectively, the main criteria were the technical, architectural, and managerial factors [72]. Saffarinia et al., defined a multi-criteria decision-making problem for the safe storage of chemicals. They proposed a new method to evaluate the severity of the incompatibility of dangerous goods based on fuzzy TOPSIS [73]. Rostamabadi et al., developed a dynamic safety analysis for process systems by using Bayesian network (BN) with Fuzzy best worst method (Fuzzy-BWM). The results showed developed Fuzzy-BWM approach can detect risks, that the BT and BN approaches weren't to detect them, and is more reliable [74].

Fu et al., used VIKOR in connection with the type-2 intuitionistic fuzzy set theory for the railway train operational risk assessment [75]. Gul et al., incorporated FAHP and FVIKOR to weigh risk parameters (i.e.: probability and severity) and to sort hazards respectively [76]. Yazdi developed a mathematical programming MCDM technique where MULTIMORA and Choquet integral under subjective circumstances are merged to rank corrective actions for an automatic fire sprinkler system in power stations. They prioritized corrective actions in a quantitative risk assessment by using an extension of DEMATEL techniques (Decision Making Trial and Evaluation Laboratory) that take into account several characteristics such as their effect on root event probability, inter-relationships, and direct/indirect costs [77]. Celik and Cebi integrated Buckley's FAHP with HFACS framework by delivering an analytical foundation and group decision-making ability to recognize the role of human errors in shipping accidents. They carried out a case study for a casualty consideration report from a bulk carrier [78].

La Fata et al., developed a methodology based on MCDM to prioritize occupational risks. To overcome the shortcomings of the traditional risk matrix, they consider the contribution of human factors to the occurrence of accidents, and a combination of Human Error Assessment and Reduction Technique (HEART) and Standardized Plant Analysis Risk Human (SPAR-H) techniques that weighted by the Analytic Hierarchy Process (AHP) method [79].

4.9 An Application of Fuzzy MCDM for Evaluation of RCMs

In past studies, RCMs were evaluated according to a variety of criteria or methods. Most of the suggested methods were designed for specific industries or risk assessment techniques and also require experts familiar with the applied methods (e.g., AHP, fuzzy logic, Bayesian networks). Also, the high number of elements that influence the choice of risk control measures increases the possibility of making mistakes in evaluations and the selection of appropriate measures, especially if it is done in a non-systematic way. In this situation, engineers and decision makers may not be able to make the right mental trade-offs when choosing a control measure. For example, the most effective risk control measure often is not also the cheapest or most practical one. Therefore, selecting the optimal measure from amongst a couple of recommended control measures is a multi-criteria decision-making problem. By structuring the criteria that influence the selection of the control measures and weighing them, we can help the professionals to systematically evaluate the recommended measures and make better decisions. Also, a generic model to evaluate the RCMs will help the users of all risk assessment techniques to make better decisions about suggested control measures. In the continuation of this chapter, first, a conceptual framework of influential elements in evaluating control measures is presented. Then, these elements were structured in a decision tree and were weighted using the Fuzzy Analytical Hierarchy Process (FAHP) approach. Finally, the proposed methods were applied to assess the recommended actions of a risk assessment project.

4.10 A Conceptual Framework for Evaluation of RCMs

Based on an initial idea of authors and literature review, a conceptual framework for evaluating RCMs is presented (Fig. 4.6). Three groups of factors influence the overall score of each control measure, which are risk-related factors (severity, probability, detectability, and exposure), quality factors (applicability, usability, reliability, duration, cost, and control strategy), and Haddon matrix components (Pr-H, Pr-M, Pr-P, Pr-S, E–H, E-M, E-P, E-S, Po-H, Po-M, Po-P, Po-S). These groups are explained below.

Risk-related factors refer to basic variables which are used to calculate or estimate a risk. According to Gul and Guneri risk is defined as the multiplication of the severity of a consequence and its probability to happen [10]. However, risk elements are not limited to the severity and probability variables. The Fine Kinney method calculates risk from the probability, exposure, and severity parameters [80]. In the FMEA method (Risk Priority Number) according to Sheahan et al. and STANDART risk is also defined by the severity, probability, and detectability parameters [81, 82]. In all of the aforementioned methods, the risk is obtained by multiplying risk parameters by each other to calculate risk scores and rank hazards. RCMs are then suggested

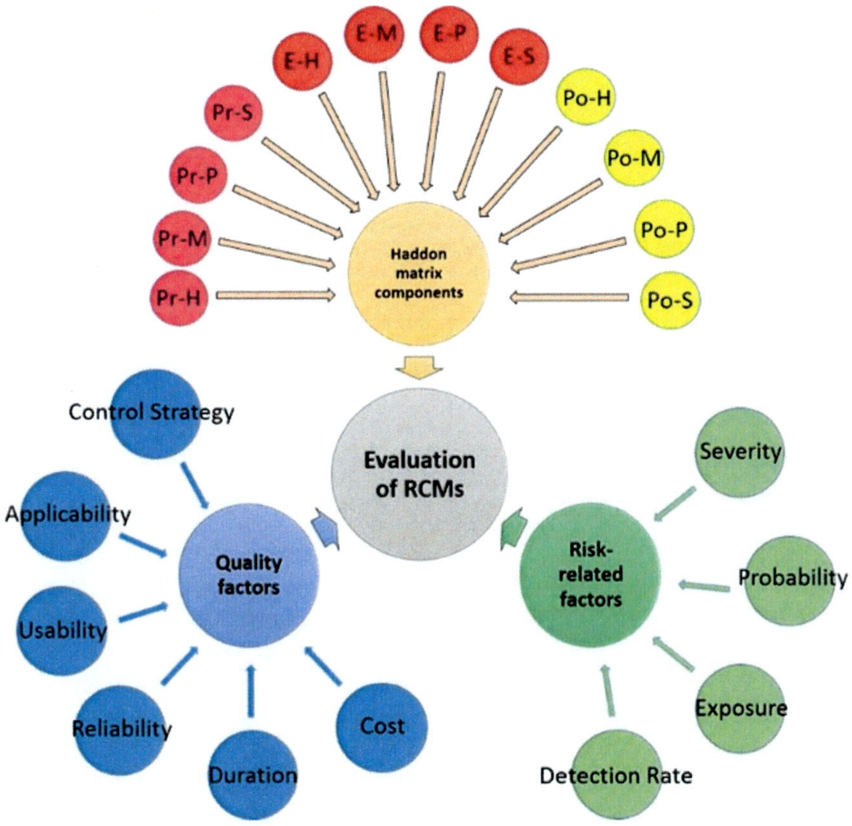

Fig. 4.6 Conceptual framework for evaluation of RCMs

for risk reduction, but it should be noted that most RCMs have a controlling effect only on some risk parameters. This effect is defined as the "effectiveness criterion" in this chapter.

Quality factors are those that influence the quality of implementation or selection of an RCM. In addition to effectiveness criterion, cost, reliability, duration, usability, applicability, and control strategy are also important criteria for selecting RCMs [83]. These criteria are defined respectively as follows: the required monetary budget to implement an RCM; the probability of correct system functioning under the predefined condition and duration; the required time to implement an RCM; the extent to which an RCM can be used by the system to achieve the risk reduction or hazard elimination and providing user satisfaction; the organization's ability to implement an RCM on existing technical and management systems and adapt the RCM to existing devices, technologies and procedures; and finally the control strategy classifies RCMs to inherently safer design (ISD), passive strategy, active strategy, procedural strategy and personal protective equipment groups.

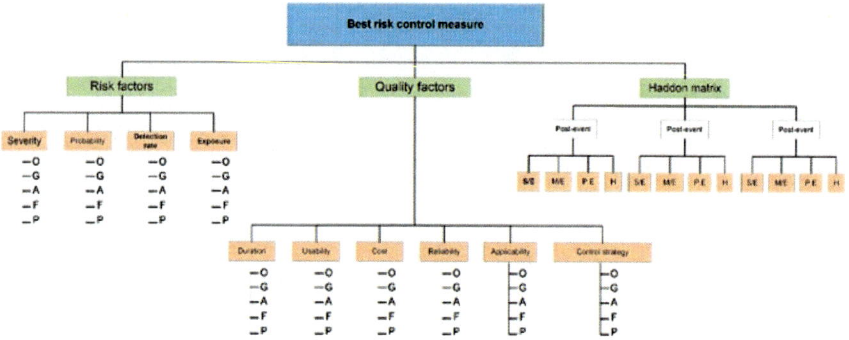

Fig. 4.7 The hierarchy decision tree for evaluating RCMs

Finally, Haddon matrix components show the time phase of implementing an RCM and the element that every RCM applies to.

A decision tree (Fig. 4.7) was created based on the conceptual framework. Then, the local and global weight of each criterion was determined through the Fuzzy Analytic Hierarch Process (FAHP) in an absolute approach. According to Liberatore et al. (1992), in absolute AHP, only the weights of alternatives (RCMs) are calculated using rating scales [68]. In this research, the rating scale suggested by Liberatore et al. (1992) was used to weight RCMs. This five-point rating scale consists of "Outstanding" (O), "Good" (G), "Average" (A), "Fair" (F), and "Poor" (P). The local weights of the rating scale were 0.53, 0.24, 0.13, 0.06, and 0.04 respectively. Liberatore et al. (1992) calculated these weights by using pairwise comparisons according to the intrinsic importance of each word.

4.11 General Technique for Evaluating Control Measures (GTECM)

A new technique proposed for the evaluation of RCMs aimed to be as simple as possible so that it can be implemented by the majority of safety professionals in various industries. We named the proposed technique the "General Technique for Evaluating Control Measures" abbreviated as GTECM. GTECM is applied to evaluating a large number of RCMs and since the number of RCMs might differ from one condition to another, an absolute AHP method is used to decrease pairwise comparison in the linguistic evaluation of each RCM. To calculate the relative weights of each criterion, the Buckley fuzzy AHP has been used. Fuzzy logic has been applied to transform the vagueness of human subjective feeling and its decision-making ability into a mathematical formula [10]. Figure 4.7 shows the decision tree constructed for this method.

The pairwise comparisons of criteria were carried out by 9 experts with working experience in occupational safety (5 experts with Ph.D. degrees in occupational safety with at least 7 years working experience, 2 experts with master's degree in industrial safety one with 6.2 and one person with 7.4 years working experience and 2 technical assistants – one with 8.2 and the other with 7.6 years working experience in implementing technical aspects of RCMs). Criteria weights are constant in the suggested method (used in equations) and only the weights of RCMs (alternatives) are variable. The steps of the developed method for evaluating and prioritizing RCMs are provided below.

4.11.1 Weighting Haddon Matrix with FAHP

The Haddon matrix classifies RCMs according to the phase time and the affected factor. Each of the 12 Haddon matrix components are composed of the junction of phase time and agent (e.g.: post-event-human). To calculate the weight of each 12 components of the Haddon matrix, one cannot compare each of the 12 components together since in the AHP method only one element can be compared to another at the same time (i.e.: it is not feasible to compare post-event-human to pre-event-human). To circumvent this, a breakdown of the matrix to its elements is applied. By using the Buckley FAHP method, the overall weight of each alternative (i.e., affected factor) constitutes the weight of each matrix component. These weights were constant and are used in the development of the RCMs evaluation equations.

To weigh RCMs according to the Haddon matrix, safety professionals will choose a component of the matrix in accordance with the feature of each RCM (e.g., for the replacement of equipment he/she will choose pre-event-machine/equipment (as shown in Table 4.21.

4.11.2 Weighing Quality and Risk-Related Factors with FAHP

Many studies aimed to determine the priority of risk parameters to assess risks [51, 84]. This research, according to the best knowledge of the authors, is the first time an evaluation of RCMs effectiveness was done according to their effect on severity, probability, exposure, and detection rate parameters. The Buckley FAHP method is used to determine the importance of each parameter for selecting the best RCM. The weight of each quality-related element is calculated in the same way using the Buckley FAHP method. The weights obtained in this step were constant and thus used in RCMs evaluation equations.

Table 4.2 Triangular fuzzy conversion scale for pairwise comparisons in the FAHP [44]

Linguistic scale	Triangular fuzzy scale	Triangular fuzzy reciprocal scale
Equally important	(1,1,1)	(1,1,1)
Intermediate value	(1,2,3)	(1/3,1/2,1)
Moderately more important	(2,3,4)	(1/4,1/3,1/2)
Intermediate value	(3,4,5)	(1/5,1/4,1/3)
Strongly more important	(4,5,6)	(1/6,1/5,1/4)
Intermediate value	(5,6,7)	(1/7,1/6,1/5)
Very strongly more important	(6,7,8)	(1/8,1/7,1/6)
Intermediate value	(7,8,9)	(1/9,1/8,1/7)
Extremely more important	(8,9,9)	(1/9,1/9,1/8)

4.11.3 Rating RCMs with Absolute AHP

To rate each RCM according to risk and quality factors, safety experts evaluate each RCM by selecting one of the linguistic variables of each criterion and the subsequent five-point rating scale. Tables 4.2 through 4.11 show the definition of the linguistic variables of each criterion according to a five-point rating scale obtained from the relevant literature [15, 80, 82, 83, 85–91]. For example, if the RCM results in an average hazard detection rate the safety expert will choose "Average" (0.13) from Table 4.5 and enter that in Table 4.21.

4.12 Results of the Overall Score

After determining the weights of each RCM using the FAHP and absolute AHP methods, and the different criteria listed in Tables 4.2 through 4.11, the results were entered into the equations and the final score for each RCM was calculated. Equations were obtained using local (RCMs) and constant (risk and quality factors) weights according to the FAHP and absolute AHP methods, meaning that they were obtained by multiplying the normalized weights of criteria by the sub-criteria and obtained results by alternatives (RCMs) weights. This is because the weights of the alternatives were obtained from the rating scales and various assessed alternatives and the weights obtained for each alternative according to different criteria were defined as a variable in the equations. The overall weights of the Haddon matrix components were constant and calculated only using the FAHP method.

Table 4.3 The linguistic variables for the severity parameter

Rating	Definition
Outstanding	A risk control measure that reduces the consequence of an accident to one or more of the following: injuries requiring first-aid measures, no loss of workdays, negligible monetary loss, and minor environmental impact or causing no concern in public opinion
Good	A risk control measure that reduces the consequence of an accident to one or more of the following: injuries that require medical services, loss of one or more workday(s), mild monetary loss, reversible mild environmental impact or causing concern in public opinion at the municipal level
Average	A risk control measure that reduces the consequence of an accident to one or more of the following: injuries resulting in permanent partial disability or hospitalization of personnel, notable monetary loss, notable yet reversible environmental impact or causing concern in public opinion at the state level
Fair	A risk control measure that reduces the consequence of an accident to one or more of the following: injury or occupational illness resulting in permanent total disability, significant monetary loss, reversible significant environmental impact or causing concern in public opinion at the federal level
Poor	A risk control measure that does not have a tangible effect on the consequence of an accident and even with its implementation one or more of the following is expected to occur: death, severe monetary loss, irreversible significant environmental impact or causing concern in public opinion at the international level

Table 4.4 The linguistic variables for the probability parameter

Rating	Definition
Outstanding	A risk control measure that once implemented, accident occurrence is assumed not to occur in the life of a system
Good	A risk control measure that once implemented, accident occurrence is assumed unlikely, but reasonably expected to happen in the life of a system
Average	A risk control measure that once implemented, accident occurrence is assumed likely in the life of a system
Fair	A risk control measure that once implemented, accident occurrence is assumed to occur several times in the life of a system
Poor	A risk control measure that does not have a tangible effect on the probability of an accident and even with its implementation could occur often in the life of a system

4.13 Classification of the Overall Scores

Overall scores obtained from the last step are classified into clusters to prioritize the RCMs. To determine the range of the overall scores in each cluster, the largest and smallest values of the overall scores were calculated. The largest value was calculated by assigning "outstanding" to all the risk and quality elements and choosing the most important components from the Haddon matrix. Since the author is not aware of any RCM that could have a control effect on all risk parameters, the least important risk

Table 4.5 The linguistic variables for the exposure parameter

Rating	Definition
Outstanding	A risk control measure that reduces the frequency of exposure to hazards to once a year
Good	A risk control measure that reduces the frequency of exposure to hazards to several times in a year
Average	A risk control measure that reduces the frequency of exposure to hazards to once a month
Fair	A risk control measure that reduces the frequency of exposure to hazards to once a week
Poor	A risk control measure that does not have a tangible effect on the exposure to hazards and even with its implementation, exposure occurs daily

Table 4.6 The linguistic variables for the detectability parameter

Rating	Definition
Outstanding	A risk control measure that, with its implementation, assumes a very high level of hazard detectability
Good	A risk control measure that with its implementation, assumes a high level of hazard detectability
Average	A risk control measure that with its implementation, assumes an average level of hazard detectability
Fair	A risk control measure that with its implementation, assumes a low level of hazard detectability
Poor	A risk control measure that does not have a tangible effect on the detectability and even with its implementation assumes a very low level of hazard detectability

Table 4.7 The linguistic variables for the cost element

Rating	Definition
Outstanding	The monetary budget required to implement the risk control measure is quite affordable and acceptable
Good	The monetary budget required to implement the risk control measure is almost affordable and acceptable
Average	The monetary budget required to implement the risk control measure is slightly affordable and acceptable
Fair	The monetary budget required to implement the risk control measure is not acceptable
Poor	The monetary budget required to implement the risk control measure is not acceptable at all

Table 4.8 The linguistic variables for the duration element

Rating	Definition
Outstanding	Duration for implementing the risk control measure is quite acceptable
Good	Duration for implementing the risk control measure is almost acceptable
Average	Duration for implementing the risk control measure is slightly acceptable
Fair	Duration for implementing the risk control measure is not acceptable
Poor	Duration for implementing the risk control measure is not acceptable at all

Table 4.9 The linguistic variables for the reliability element

Rating	Definition
Outstanding	The probability of the risk control measure functioning correctly under predefined conditions and duration is very high
Good	The probability of the risk control measure functioning correctly under predefined conditions and duration is high
Average	The probability of the risk control measure functioning correctly under predefined conditions and duration is average
Fair	The probability of the risk control measure functioning correctly under predefined conditions and duration is low
Poor	The probability of the risk control measure functioning correctly under predefined conditions and duration is very low

Table 4.10 The linguistic variables for the usability element

Rating	Definition
Outstanding	The use of this risk control measure is quite acceptable at providing user satisfaction and controlling risk
Good	The use of this risk control measure is almost acceptable at providing user satisfaction and controlling risk
Average	The use of this risk control measure is slightly acceptable at providing user satisfaction and controlling risk
Fair	The use of this risk control measure is not acceptable at providing user satisfaction and controlling risk
Poor	The use of this risk control measure is never acceptable at providing user satisfaction and controlling risk

parameters were considered as "poor" to calculate the largest value. The risk element with the smallest value has been chosen to make sure that all the possible overall scores are smaller than the calculated value. The smallest value was calculated by assigning "poor" to all the risk and quality elements and choosing the least important component of the Haddon matrix. The overall scores are obtained by subtracting the highest value from the lowest, dividing the outcome by 4, and continuously summing up the result to the smallest value.

Table 4.11 The linguistic variables for the applicability element

Rating	Definition
Outstanding	Implementation of a risk control measure on existing technical and management systems and its adaptation to equipment, working conditions, technology, and executive procedures is quite feasible
Good	Implementation of a risk control measure on existing technical and management systems and its adaptation to equipment, working conditions, technology, and executive procedures is almost feasible
Average	Implementation of a risk control measure on existing technical and management systems and its adaptation to equipment, working conditions, technology, and executive procedures is slightly feasible
Fair	Implementation of a risk control measure on existing technical and management systems and its adaptation to equipment, working conditions, technology, and executive procedures is not feasible
Poor	Implementation of a risk control measure on existing technical and management systems and its adaptation to equipment, working conditions, technology, and executive procedures is not feasible at all

Table 4.12 The linguistic variables for the control strategies element

Rating	Definition
Outstanding	Risk control measures that cause a significant reduction in individual or collective risk by using inherently safer design principles and changing system design
Good	Risk control measures that cause a reduction in individual or collective risk by using passive control strategies such as proper placement of equipment or materials
Average	Risk control measures that cause a reduction in individual or collective risk by using active control strategies such as sensors, safety systems, or sprinklers
Fair	Risk control measures that cause a reduction in individual or collective risk by using a procedural strategy such as operational standards, training, emergency response procedures, and management systems
Poor	Risk control measures that cause a reduction in individual risk by using personal protective equipment

Table 4.13 Group fuzzy pairwise comparison matrix of criteria in evaluating RCMs

	Haddon Matrix	Risk factor	Quality factor	Local Weight	Rank
Haddon Matrix	(1, 1, 1)	(0.24, 0.29, 0.37)	(0.38, 0.47, 0.58)	0.153	3
Risk factor	(2.68, 3.43, 4.18)	(1, 1, 1)	(1.16, 1.57, 2.07)	0.517	1
Quality factor	(1.72, 2.13, 2.65)	(0.48, 0.64, 0.86)	(1, 1, 1)	0.330	2

$\lambda_{\max}^{m} = 3.000$, $CR^{m} = 0.0001$ $\lambda_{\max}^{g} = 3.000$, $CR^{g} = 0.0000$

4.14 Application and Results

To demonstrate the applicability, simplicity and benefits of the GTECM, RCMs suggested by Pouyakian were evaluated and the proposed method was applied in the Glucosan Company (Qazvin, Iran) using ETBA hazard identification and MIL-STD-882B risk assessment methods [92]. This company is the largest manufacturer of glucose, starch, gluten, and corn oil in Iran. Unloading and storing corn, corn grain, steep, starching, glucose storage and packaging, and hydrochloric acid storage are the main functions of the factory.

According to Pouyakian, occupational noise-induced hearing loss was the most common risk, especially in the starch production unit, and several RCMs were suggested in this regard. These RCMs were evaluated by the newly proposed method (GTECM). Other reasons these RCMs were selected include the completeness of the information and the variation in their features. RCM evaluation was conducted by three experts (two safety experts and one technical assistant) who are experienced in occupational health and safety concerning controlling risks in industrial environments. Experts A and B had more than 8 years of experience in the occupational safety field and were responsible for suggesting and implementing RCMs. Expert C has worked in industrial safety as a technical assistant for more than 5 years and has often carried out technical aspects of RCMs.

The application and results of the GTECM as applied to occupational noise-induced hearing loss RCMs in the starching compound is as follows:

Figure 4.6 shows the importance of three criteria (Haddon matrix, risk and quality factors) for evaluating RCMs obtained through the group fuzzy pairwise comparison matrix in Table 4.12. As defuzzified and normalized weights show, risk-related factors, quality factors and Haddon matrix ranked highest in importance in the GTECM.

The weights of the time phases that the Haddon matrix gained through the group fuzzy pairwise comparison matrix are shown in Table 4.13. Calculating the importance of the time phases was vital in evaluating RCMs as detecting the phase time that an RCM applies its effect could have great importance on controlling losses. As predicted, the pre-event phase, due to its prevention effect of the accident, was more important than other time phases in the RCMs evaluation process.

Groups of normalized and defuzzified weights of the affected agent in the Haddon matrix for each phase time obtained by the Buckley FAHP method are shown in Tables 4.14 through 4.16. These weights determine which agent has a higher value of the control. The results show that the "machine/equipment" element was most important according to the different time phases which may be due to the higher reliability of machine/equipment as compared to human and/or socio/economic factors and also less importance of the physical environment in controlling risks when dealing with accidents. Stout et al. attribute this to the fact that humans have low reliability and an unpredictable response when dealing with accidents and crises [93]. These results show that the RCMs that have a control effect on machines/equipment have top priority.

The normalized and weighted Haddon matrix has been calculated by multiplying defuzzified and normalized weights of each agent in its upper time phase. The weighted Haddon matrix is shown in Table 4.17. The results demonstrate that the pre-event-machine/equipment component of the matrix bears the most weight. Also, according to AHP logic, the sum of all Haddon matrix components is equal to 1, which verifies the calculations.

To calculate the score of the Haddon matrix component (HM) for each RCM considering the weights in Table 4.17 and the importance of the Haddon matrix criterion in Table 4.12 the following equation has been applied, where 'Wh' represents the weight of the Haddon matrix component chosen from Table 4.17 according to RCM features.

$$\mathbf{HM} = \mathbf{Wh} * 0.153 \tag{4.13}$$

In the case study, a component of the Haddon matrix was chosen by experts according to the characteristics of each RCM and its weight was entered into Eq. 4.13. Based on Table 4.21, "Installing sound-absorbing materials and acoustic tiles in the hall" is a pre-event physical environment component of the Haddon matrix with a weight (according to Table 4.17) of 0.159. By entering the obtained weight into Eq. 4.13, a Haddon matrix score (HM) of 0.024 was obtained for the RCM chosen. Table 4.21 contains the evaluation results of other RCMs according to the Haddon matrix.

Risk-related factors are the next criteria for evaluating RCMs. Table 4.18 illustrated the group fuzzy pairwise comparison matrix for all risk elements by using Buckley FAHP method. According to Table 4.18, respectably severity, detection rate, probability, and exposure parameters were the most important risk-related factors in evaluating RCMs. The reason for applying the defuzzified weights of risk parameters to the GTECM is that the importance of each risk element in evaluating RCMs is different and the final score of RCMs is computed more precisely (Table 4.15).

As mentioned, the GTECM uses rating scales with linguistic variables for evaluating various RCMs according to risk elements. The importance of each RCM is obtained Tables 4.2, 4.3, 4.4 and 4.5 based on various risk parameters. Therefore, RCMs weights will vary due to their different controlling effect on each risk element. In order to calculate the risk-related factor score (RF) for each RCM according to the

Table 4.14 Group fuzzy pairwise comparison matrix of time phases of Haddon matrix

	After Accident	During Accident	Before Accident	Local Weight	Rank
After Accident	(1, 1, 1)	(0.19, 0.22, 0.27)	(0.12, 0.12, 0.14)	0.069	3
During Accident	(3.69, 4.49, 5.39)	(1, 1, 1)	(0.22, 0.28, 0.40)	0.253	2
Before Accident	(6.37, 7.39, 8.64)	(2.49, 3.54, 4.57)	(1, 1, 1)	0.678	1
	$\lambda_{\max}^{m} = 3.0144$, $\mathrm{CR}^{\mathrm{m}} = 0.0147$		$\lambda_{\max}^{g} = 3.0264$, $\mathrm{CR}^{\mathrm{g}} = 0.0735$		

Table 4.15 Group fuzzy pairwise comparison matrix of the affected agent in Haddon matrix concerning pre-event time phase

	Human	Physical Environment	Machine/Equipment	Social/Economic Factors	Local Weight	Rank
Human	(1, 1, 1)	(0.71, 0.84, 0.99)	(0.65, 0.78, 0.93)	(0.92, 1.13, 1.36)	0.229	4
Physical Environment	(1.01, 1.20, 1.42)	(1, 1, 1)	(0.51, 0.67, 0.93)	(0.76, 0.93, 1.16)	0.234	3
Machine/Equipment	(1.07, 1.28, 1.54)	(1.07, 1.49, 1.94)	(1, 1, 1)	(0.88, 1.04, 1.25)	0.295	1
Social Economic Factors	(0.74, 0.89, 1.08)	(0.86, 1.07, 1.32)	(0.80, 0.96, 1.14)	(1, 1, 1)	0.242	2
	$\lambda^{m}_{\max} = 4.0273$, $CR^{m} = 0.0115$		$\lambda^{g}_{\max} = 4.0237$, $CR^{g} = 0.030$			

Table 4.16 Group fuzzy pairwise comparison matrix of the affected agent in Haddon matrix concerning event time phase

	Human	Physical Environment	Machine/Equipment	Social Economic Factors	Local Weight	Rank
Human	(1, 1, 1)	(0.47, 0.55, 0.65)	(0.46, 0.52, 0.60)	(0.63, 0.77, 0.96)	0.166	4
Physical Environment	(1.53, 1.82, 2.11)	(1, 1, 1)	(0.73, 0.91, 1.12)	(1.14, 1.53, 1.95)	0.303	2
Machine/Equipment	(1.65, 1.91, 2.15)	(0.90, 1.10, 1.37)	(1, 1, 1)	(0.99, 1.32, 1.69)	0.311	1
Social Economic Factors	(1.04, 1.30, 1.58)	(0.51, 0.65, 0.88)	(0.59, 0.76, 1.01)	(1, 1, 1)	0.220	3
	$\lambda^{m}_{max} = 4.0049$, $CR^{m} = 0.0021$		$\lambda^{g}_{max} = 4.0052$, $CR^{g} = 0.0066$			

Table 4.17 Group fuzzy pairwise comparison matrix of the affected agent in Haddon matrix concerning post-event time phase

	Human	Physical environment	Machine/Equipment	Social economic factors	Local weight	Rank
Human	(1, 1, 1)	(1.27, 1.51, 1.79)	(0.73, 0.89, 1.10)	(0.92, 1.10, 1.34)	0.272	2
Physical environment	(0.56, 0.66, 0.79)	(1, 1, 1)	(0.44, 0.58, 0.79)	(0.52, 0.65, 0.86)	0.177	4
Machine/Equipment	(0.91, 1.12, 1.37)	(1.26, 1.71, 2.25)	(1, 1, 1)	(0.86, 1.08, 1.37)	0.297	1
Social economic factors	(0.75, 0.91, 1.09)	(1.04, 1.34, 1.72)	(0.73, 0.92, 1.17)	(1, 1, 1)	0.254	3
	$\lambda^{m}_{max} = 4.0315$, $CR^{m} = 0.0132$		$\lambda^{g}_{max} = 4.0246$, $CR^{g} = 0.0313$			

Table 4.18 The weighted Haddon matrix

Haddon matrix	Human	Physical environment	Machine/equipment	Social/economic factors
Pre-event	0.155	0.159	0.200	0.164
Event	0.042	0.077	0.079	0.056
Post-event	0.019	0.012	0.020	0.017

defuzzified and normalized weights of risk parameters (Table 4.18), the importance of risk factor criterion (Table 4.12), and the absolute AHP method, the following equation is used:

$$\mathbf{RF} = [(0.374 * \mathbf{WS}) + (0.211 * \mathbf{WP}) + (0.225 * \mathbf{WD}) + (0.190 * \mathbf{WE})] * 0.517 \tag{4.14}$$

In that:

WS : The selected weight from the five-point rating scale was chosen according to the severity criterion from Table 4.2.
WP : The selected weight from the five-point rating scale was chosen according to the probability criterion from Table 4.3.
WD : The selected weight from the five-point rating scale was chosen according to the detection rate criterion from Table 4.4.
WE : The selected weight from the five-point rating scale was chosen according to the exposure criterion from Table 4.5.

The experts in the case study believed that the RCM "Installing sound-absorbing materials and acoustic tiles in the hall" did not have any effect on the detection rate parameter concerning Table 4.5. However, according to Table 4.2 through 4.4, the RCM had a good effect on severity, probability and the average effect on exposure elements. To calculate the risk-related factors score, the weight of each five-point rating scale was chosen according to each criterion inserted in Eq. 4.14. This means that they assigned 0.24 for severity and probability, 0.04 for detection rate and 0.13 for exposure parameters. Finally, the risk-related factors score (RF) achieved was 0.09. Table 4.21 shows the evaluation results of the other RCMs according to risk-related factors.

The quality factor is the last criterion for evaluating RCMs. Table 4.19 shows the group fuzzy pairwise comparison matrix of each quality element as expressed in risk parameters where reliability, cost, usability, control strategy, applicability and

Table 4.19 Group fuzzy pairwise comparison matrix of risk parameters

	Severity	Probability	Exposure	Detection rate	Local Weight	Rank
Severity	(1, 1, 1)	(1.46, 1.99, 2.57)	(1.75, 2.20, 2.71)	(1.07, 1.34, 1.69)	0.374	1
Probability	(0.39, 0.50, 0.68)	(1, 1, 1)	(0.79, 1.03, 1.35)	(0.87, 1.09, 1.36)	0.211	3
Exposure	(0.37, 0.45, 0.57)	(0.74, 0.97, 1.27)	(1, 1, 1)	(0.75, 0.88, 1.04)	0.190	4
Detection Rate	(0.59, 0.75, 0.94)	(0.74, 0.91, 1.15)	(0.96, 1.14, 1.33)	(1, 1, 1)	0.225	2
	$\lambda^{m}_{max} = 4.0261$, $CR^{m} = 0.0110$		$\lambda^{g}_{max} = 4.0231$, $CR^{g} = 0.029$			

duration factors were the most important quality elements in evaluating RCMs. The reason for prioritizing the quality elements was to show the importance of reliability, cost, usability, control strategy, applicability and duration elements in selecting the most appropriate RCM and to illustrate that the cost and risk elements are not the only factors for selecting the best RCM. Although the results show that the cost and risk elements are the most important factors in evaluating RCMs.

To evaluate RCMs according to quality-related factors as mentioned before, the GTECM uses a rating scale with linguistic variables. The importance of each RCM was obtained according to different quality elements using Table 4.6 through 4.11. The obtained weights for RCMs are variable concerning quality elements as in risk parameters. The quality-related factors score (QF) for each RCM is calculated using Eq. 4.15 according to the defuzzified and normalized weights of quality elements (Table 4.19), the importance of quality factor criterion (Table 4.12), and the absolute AHP method as per the following:

$$\mathbf{QF} = [(0.161 * \mathbf{WA}) + (0.199 * \mathbf{WR}) + (0.191 * \mathbf{WC}) + (0.169 * \mathbf{WSt}) + (0.183 * \mathbf{WU}) + (0.097 * \mathbf{WDu}) * 0.330 \tag{4.15}$$

W_A : The selected weight from the five-point rating scale was chosen according to the applicability criterion in Table 4.10.

W_R : The selected weight from the five-point rating scale was chosen according to the reliability criterion in Table 4.8.

W_C : The selected weight from the five-point rating scale chosen according to the cost criterion in Table 4.6.

W_{St} : The selected weight from the five-point rating scale was chosen according to the control strategy criterion in Table 4.11.

W_U : The selected weight from the five-point rating scale was chosen according to the usability criterion in Table 4.9.

W_{Du} : The selected weight from the five-point rating scale was chosen according to the duration criterion in Table 4.7.

The results from the evaluation of the RCM "Installing sound-absorbing materials and acoustic tiles in the hall", according to quality-related factors concerning Table 4.6 through 4.11 are "average" (0.13) for applicability, "good" (0.24) for control strategy and usability, "fair" (0.06) for cost and duration and "outstanding" for reliability. By applying these weights to Eq. 4.16, the QF is equal to 0.075. Table 4.21 illustrates the results of evaluating other RCMs according to quality-related factors.

The final score of evaluating each RCM obtained by summing the Haddon matrix (HM), risk-related factors (RF) and quality-related factors (QF) scores are as follows (Eq. 4.16):

$$\mathbf{OS} = (\mathbf{RF} + \mathbf{QF} + \mathbf{HM}) * 100 \tag{4.16}$$

Using the Eq. (4.16), the overall score (OS) for the RCM "Installing sound-absorbing materials and acoustic tiles in the hall" is equal to 18.9.

The RCMs are then classified according to their overall scores. The largest and smallest possible overall scores concerning the exposure parameter which had the least importance in risk elements were 43.1 and 3.6 respectably. Table 4.20 shows the proposed four-group classification of the overall scores for ranking RCMs, where the RCMs with an overall score of "outstanding", "good", "average" and "poor" are the top priority for implementation. Within the "average" cluster, the top priority is given to the RCMs with the largest scores. The particular definitions of the overall scores and their ranges are general and may be subject to change based on policies, organizational conditions, and the overall risk level of the system, such as the William Fine risk assessment method [27].

The evaluation of the RCMs suggested for controlling the risk of occupational noise-induced hearing loss reveals that most RCMs achieve an "average" score. The results also show that based on Table 4.21, replacing noisy equipment with quieter ones, identifying and repairing broken components and fittings to reduce noise, constructing a noise-resistant operator's room, training for proper earmuff use, installing appropriate isolators on equipment for reduced noise and vibration, installing sound-absorbing materials and acoustic tiles in the hall and selecting and providing appropriate earmuffs for workers were top priority. As expected, replacing noisy equipment with quieter ones was the only RCM that achieved an "outstanding" classification since it can completely eliminate the hazard (refer to Table 4.22).

4.15 Conclusion

Evaluating and ranking the RCMs at the process of risk assessment methods have got little attention in the previous studies. Most of the studies have not paid attention to this critical part of the risk assessment and management process [22, 23, 94]. Other studies have prioritized the RCMs in a qualitative manner and without any systematic approach [9, 10]. The purpose of this study was to present a new framework for evaluating RCMs, considering the weight of each criterion on the final decision and providing a structured method without the need to complicated computations for users. We combined the strengths of three Risk Control Models (Haddon Matrix, Hierarchy of Controls, and Chemical Process Safety Strategies) to develop the new method.

A review of most hazard identification and risk assessment studies shows that suggested RCMs are seldom evaluated using a theoretical method and in the majority of cases they are merely classified verbally according to different hazard control hierarchies. The use of incomplete criteria and a lack of a theoretical approach for evaluating RCMs means that the solutions offered are often limited and suggested RCMs are not prioritized. AHP method, despite being an old method, has been used in a recent study to develop a risk evaluation method [51]. This is because of its strong logical structure to formulate decision-making problems in the risk assessment context.

Table 4.20 Group fuzzy pairwise comparison matrix of quality elements

	Control strategies	Applicability	Reliability	Cost	Usability	Duration	Local weight	Rank
Control strategies	(1, 1, 1)	(0.71, 0.85, 1.03)	(0.60, 0.75, 0.93)	(0.61, 0.80, 1.11)	(0.94, 1.18, 1.46)	(1.71, 2.13, 2.68)	0.169	4
Applicability	(0.97, 1.18, 1.41)	(1, 1, 1)	(0.63, 0.71, 0.82)	(0.61, 0.83, 1.13)	(0.66, 0.79, 0.96)	(1.32, 1.78, 2.32)	0.161	5
Reliability	(1.08, 1.34, 1.67)	(1.21, 1.41, 1.58)	(1, 1, 1)	(0.76, 0.99, 1.35)	(0.90, 1.08, 1.28)	(1.36, 1.77, 2.27)	0.199	1
Cost	(0.90, 1.25, 1.63)	(0.89, 1.21, 1.65)	(0.74, 1.01, 1.31)	(1, 1, 1)	(0.68, 0.88, 1.20)	(1.34, 1.88, 2.50)	0.191	2
Usability	(0.69, 0.85, 1.06)	(1.04, 1.27, 1.51)	(0.78, 0.93, 1.12)	(0.85, 1.17, 1.53)	(1, 1, 1)	(1.53, 1.84, 2.14)	0.183	3
Duration	(0.37, 0.47, 0.58)	(0.43, 0.56, 0.76)	(0.44, 0.56, 0.74)	(0.40,0.53, 0.74)	(0.47, 0.54, 0.65)	(1, 1, 1)	0.097	6
	$\lambda^{m}_{\max} = 6.0498$, $\mathrm{CR}^{\mathrm{m}} = 0.0083$			$\lambda^{g}_{\max} = 6.0459$, $\mathrm{CR}^{g} = 0.0241$				

Table 4.21 The RCMs overall score definitions

Overall score	RCM ranking
33.3–43.1	Outstanding
23.5–33.2	Good
13.6–23.4	Average
3.6–13.5	Poor

Considering the fact that implementing an inappropriate RCM concerning organizational resource constraints is a big problem, and to fill this gap and evaluate various RCMs comprehensively, a novel solution based on the FAHP and absolute AHP method, using a complete set of criteria, is proposed in this paper.

For the evaluation of the effects of the suggested RCMs on risk-related factors, quality-related factors, and Haddon matrix components, the Buckley FAHP method was used. To simplify calculations and remove the need for a pairwise comparisons matrix for each RCM, a five-point rating scale based on linguistic variables was developed using the absolute AHP method for each criterion (risk and quality factors). The linguistic variables were defined as comprehensively as possible for better applicability in various industries. For ranking, a four-group classification system of the overall score of each RCM was devised. Because the AHP method is used in designing the GTECM, even RCMs that fall in the same class can be prioritized according to the magnitude of their overall score. A case study for evaluating suggested RCMs for occupational noise-induced hearing loss risk was used to illustrate the applicability, simplicity, and benefits of the GTECM. Using a complete set of risk parameters enables the GTECM to be applicable for evaluating the majority of suggested RCMs in the various hazard identification and risk assessment techniques.

Applying the GTECM on RCMs enables the simple and precise scoring, classifying, and prioritizing of suggested RCMs without the need for pairwise comparisons and fuzzy logic computations. It also makes it possible for the occupational safety and health experts to determine which dimensions of the Haddon matrix, risk, and quality factors the suggested RCMs have a control effect on. In addition, the obtained weights can clarify which of the criteria are more important in the selection of the best RCM.

Further studies would also widen the applicability of the proposed method. Due to some limitations of the study, the relationship between the quality and risk factor sub-criteria, developing fuzzy sets on absolute AHP, calculating exclusive weights for each sub-criteria in the five-point rating scales, and using a new MCDM technique or a new method for ranking fuzzy numbers to extend the GTECM are suggestions for future studies.

Table 4.22 GTECM worksheet for evaluation of suggested RCMs for reducing risk of occupational noise-induced hearing

Risk control measure	Evaluation of risk control measures											Overall Score	Classification
	Risk-related factors				Quality-related factors						Haddon matrix component		
	Severity	Probability	Detection rate	Exposure	Applicability	Control strategy	Cost	Duration	Reliability	Usability			
Selecting and providing appropriate earmuffs for workers	Good	Good	Poor	Poor	Outstanding	Poor	Good	Outstanding	Average	Average	Pre-event + Human	18.4	Average
Training for proper earmuff use	Good	Average	Average	Poor	Outstanding	Fair	Outstanding	Good	Fair	Good	Pre-event + Human	19.4	Average
Identifying and repairing broken components and fittings to reduce noise	Good	Good	Poor	Good	Outstanding	Good	Good	Average	Outstanding	Outstanding	Pre-event + Machine/equipment	25.9	Good
Installing appropriate isolators on equipment for reduced noise and vibration	Good	Good	Poor	Good	Average	Good	Average	Fair	Good	Good	Pre-event + Machine/equipment	19.2	Average
Replacing noisy equipment with quieter ones	Outstanding	Outstanding	Poor	Outstanding	Poor	Outstanding	Poor	Poor	Outstanding	Outstanding	Pre-event + Machine/equipment	35	Outstanding
Constructing a noise resistant operator's room	Outstanding	Good	Poor	Average	Good	Good	Average	Fair	Outstanding	Average	Pre-event + Physical environment	24.9	Good

(continued)

Table 4.22 (continued)

Risk control measure	Evaluation of risk control measures											Overall Score	Classification
	Risk-related factors				Quality-related factors						Haddon matrix component		
	Severity	Probability	Detection rate	Exposure	Applicability	Control strategy	Cost	Duration	Reliability	Usability			
Installing sound absorbing materials and acoustic tiles in the hall	Good	Good	Poor	Average	Average	Good	Fair	Fair	Outstanding	Good	Pre-event + Physical environment	18.9	Average

References

1. P.K. Marhavilas, D. Koulouriotis, V. Gemeni, Risk analysis and assessment methodologies in the work sites: on a review, classification and comparative study of the scientific literature of the period 2000–2009. J. Loss Prev. Process Ind. **24**(5), 477–523 (2011)
2. X. Ning, J. Qi, C. Wu, A quantitative safety risk assessment model for construction site layout planning. Saf. Sci. **104**, 246–259 (2018)
3. J. Tixier, G. Dusserre, O. Salvi, D. Gaston, Review of 62 risk analysis methodologies of industrial plants. J. Loss Prev. Process Ind. **15**(4), 291–303 (2002)
4. V. Villa, N. Paltrinieri, F. Khan, V. Cozzani, Towards dynamic risk analysis: a review of the risk assessment approach and its limitations in the chemical process industry. Saf. Sci. **89**, 77–93 (2016)
5. F. Yan, K. Xu, Methodology and case study of quantitative preliminary hazard analysis based on cloud model. J. Loss Prev. Process Ind. **60**, 116–124 (2019)
6. F. Zammori, R. Gabbrielli, ANP/RPN: a multi criteria evaluation of the risk priority number. Qual. Reliab. Eng. Int. **28**(1), 85–104 (2012)
7. G. Di Bona, A. Silvestri, A. Forcina, A. Petrillo, Total efficient risk priority number (TERPN): a new method for risk assessment. J. Risk Res. **21**(11), 1384–1408 (2018)
8. Ericson CA. Hazard analysis techniques for system safety: John Wiley & Sons; 2015.
9. L. Botti, V. Duraccio, M.G. Gnoni, C. Mora, An integrated holistic approach to health and safety in confined spaces. J. Loss Prev. Process Ind. **55**, 25–35 (2018)
10. M. Gul, A.F. Guneri, A fuzzy multi criteria risk assessment based on decision matrix technique: a case study for aluminum industry. J. Loss Prev. Process Ind. **40**, 89–100 (2016)
11. D.C. Hendershot, Inherently safer design: an overview of key elements. Prof. Saf. **56**(2), 48 (2011)
12. W. Haddon Jr., On the escape of tigers: an ecologic note. Am. J. Public Health Nations Health **60**(12), 2229–2234 (1970)
13. C.W. Runyan, Using the Haddon matrix: introducing the third dimension. Inj. Prev. **4**(4), 302–307 (1998)
14. D. Zhao, A.P. McCoy, B.M. Kleiner, T.L. Smith-Jackson, Control measures of electrical hazards: an analysis of construction industry. Saf. Sci. **77**, 143–151 (2015)
15. M.J. Jafari, H. Mohammadi, G. Reniers, M. Pouyakian, F. Nourai, S.A. Torabi et al., Exploring inherent process safety indicators and approaches for their estimation: a systematic review. J. Loss Prev. Process Ind. **52**, 66–80 (2018)
16. F.I. Khan, P.R. Amyotte, I2SI: a comprehensive quantitative tool for inherent safety and cost evaluation. J. Loss Prev. Process Ind. **18**(4), 310–326 (2005)
17. R.F. Wasileski, F. Henselwood, LOPA onions: Peeling back the outer layers. Process Saf. Prog. **30**(2), 122–125 (2011)
18. W. Haddon Jr (ed), Approaches to prevention of injuries. Proceedings o the American Medical Association Conference on Prevention of Disabling Injuries, Miami, FL, USA, (1983)
19. C.W. Runyan, Introduction: back to the future—revisiting Haddon's conceptualization of injury epidemiology and prevention. Epidemiol. Rev. **25**(1), 60–64 (2003)
20. M. Jafari, M. Pouyakian, S. Hanifi, Development of a framework for dynamic risk assessment of environmental impacts in chemicals warehouse using CFD-BN. Int. J. Environ. Sci. Technol., 1–16 (2021)
21. M. Pouyakian, M.J. Jafari, F. Laal, F. Nourai, E. Zarei, A comprehensive approach to analyze the risk of floating roof storage tanks. Process Saf. Environ. Prot. **146**, 811–836 (2021)
22. H.T. Liu, Y.I. Tsai, A fuzzy risk assessment approach for occupational hazards in the construction industry. Safety Sci. **50**(4), 1067–1078 (2012)
23. A. Nieto-Morote, F. Ruz-Vila, A fuzzy approach to construction project risk assessment. Int. J. Project Manage. **29**(2), 220–231 (2011)
24. M. Pouyakian, A. Khatabakhsh, M.J. Jafari, The analysis of hazard identification and risk assessment studies with the approach to assessing risk control measures since 2001 to 2017: a systemic review. Iran Occupational Health J. **16**(6), 1–15 (2020)

25. E. Zarei, M. Yazdi, R. Abbassi, F. Khan, A hybrid model for human factor analysis in process accidents: FBN-HFACS. J. Loss Prev. Process Ind. **57**, 142–155 (2019)
26. E. Vanem, Ø. Endresen, R. Skjong, Cost-effectiveness criteria for marine oil spill preventive measures. Reliab. Eng. Syst. Saf. **93**(9), 1354–1368 (2008)
27. W.T. Fine, Mathematical evaluations for controlling hazards. NAVAL ORDNANCE LAB WHITE OAK MD (1971)
28. M.J. Ellenbecker, Engineering controls as an intervention to reduce worker exposure. Am. J. Ind. Med. **29**(4), 303–307 (1996)
29. D.N. Atehnjia, Z. Yang, J. Wang, Risk control and cost benefit analysis of docking operation. IJASRE, **4** (2018)
30. S. Young, From zero to hero. A case study of industrial injury reduction: New Zealand Aluminium Smelters Limited. Safety Sci. **64**, 99–108 (2014)
31. K.V. Kortenkamp, C.F. Moore, D.P. Sheridan, E.S. Ahrens, No hiking beyond this point! hiking risk prevention recommendations in peer-reviewed literature. J. Outdoor Recreat. Tour. **20**, 67–76 (2017)
32. B.J. Maguire, B.J. O'Neill, P. O'Meara, M. Browne, M.T. Dealy, Preventing EMS workplace violence: a mixed-methods analysis of insights from assaulted medics. Injury **49**(7), 1258–1265 (2018)
33. M. Yazdi, A. Nedjati, E. Zarei, R. Abbassi, A novel extension of DEMATEL approach for probabilistic safety analysis in process systems. Saf. Sci. **121**, 119–136 (2020)
34. M. Yazdi, F. Khan, R. Abbassi, R. Rusli, Improved DEMATEL methodology for effective safety management decision-making. Saf. Sci. **127**, 104705 (2020). https://doi.org/10.1016/j.ssci.2020.104705
35. W. Ho, Integrated analytic hierarchy process and its applications–a literature review. Eur. J. Oper. Res. **186**(1), 211–228 (2008)
36. Divya C, Raju LS, Singaravel B. Application of MCDM Methods for Process Parameter Optimization in Turning Process—A Review. Recent Trends in Mechanical Engineering. 2021:199–207.
37. E.K. Zavadskas, Z. Turskis, S. Kildienė, State of art surveys of overviews on MCDM/MADM methods. Technol. Econ. Dev. Econ. **20**(1), 165–179 (2014)
38. Roy PK, Shaw K. An integrated fuzzy model for evaluation and selection of mobile banking (m-banking) applications using new fuzzy-BWM and fuzzy-TOPSIS. Complex & Intelligent Systems. 2021.
39. S. Kubler, J. Robert, W. Derigent, A. Voisin, Y. Le Traon, A state-of the-art survey and testbed of fuzzy AHP (FAHP) applications. Expert Syst. Appl. **65**, 398–422 (2016)
40. M. Gul, E. Celik, N. Aydin, A.T. Gumus, A.F. Guneri, A state of the art literature review of VIKOR and its fuzzy extensions on applications. Appl. Soft Comput. **46**, 60–89 (2016)
41. A. Mardani, A. Jusoh, K. Nor, Z. Khalifah, N. Zakwan, A. Valipour, Multiple criteria decision-making techniques and their applications–a review of the literature from 2000 to 2014. Econ. Res.-Ekonomska Istraživanja. **28**(1), 516–571 (2015)
42. N. Bhagtani, *A Better Tool for Environmental Decision Making: Comparing MCDA with CBA* (University of East Anglia, UK, 2008)
43. A. Khatabakhsh, M. Pouyakian, M. J. Jafari, Towards improving risk management: a review of risk control measures prioritization patterns and provision of a new conceptual framework. J. Health Field, **8**(3) (2020)
44. A. Emrouznejad, W. Ho, Fuzzy analytic hierarchy process, CRC Press (2017)
45. E. Fontela, A. Gabus, The DEMATEL observer. Dematel (1976)
46. T.L. Saaty, A scaling method for priorities in hierarchical structures. J. Math. Psychol. **15**(3), 234–281 (1977)
47. T.L. Saaty, Fundamentals of the analytic network process—dependence and feedback in decision-making with a single network. J. Syst. Sci. Syst. Eng. **13**(2), 129–157 (2004)
48. J. Rezaei, Best-worst multi-criteria decision-making method. Omega **53**, 49–57 (2015)
49. G.H. Tzeng, J. J. Huang, Multiple Attribute Decision Making: Methods and Applications: CRC Press (2011)

50. S. Opricovic, G.-H. Tzeng, Compromise solution by MCDM methods: a comparative analysis of VIKOR and TOPSIS. Eur. J. Oper. Res. **156**(2), 445–455 (2004)
51. R. Fattahi, M. Khalilzadeh, Risk evaluation using a novel hybrid method based on FMEA, extended MULTIMOORA, and AHP methods under fuzzy environment. Saf. Sci. **102**, 290–300 (2018)
52. M. Yazdi, Risk assessment based on novel intuitionistic fuzzy-hybrid-modified TOPSIS approach. Saf. Sci. **110**, 438–48 (2018). https://doi.org/10.1016/j.ssci.2018.03.005
53. H. Li, J.Y. Guo, M. Yazdi, A. Nedjati, K.A. Adesina, Supportive emergency decision-making model towards sustainable development with fuzzy expert system. Neural Comput Appl **33**, 15619–15637 (2021). https://doi.org/10.1007/s00521-021-06183-4
54. T. L. Saaty, The analytic hierarchy process: planning, priority setting, resource allocation: McGraw-Hill International Book Company (1980)
55. T. L. Saaty, Multicriteria decision making-the analytic hierarchy process: planning, priority setting, resource allocation: RWS publications (1990)
56. S. Kubler, A. Voisin, W. Derigent, A. Thomas, É. Rondeau, K. Främling, Group fuzzy AHP approach to embed relevant data on "communicating material." Comput. Ind. **65**(4), 675–692 (2014)
57. A. Ishizaka, A. Labib, Review of the main developments in the analytic hierarchy process. Expert Syst. Appl. **38**(11), 14336–14345 (2011)
58. Z. Xu, H. Liao, Intuitionistic fuzzy analytic hierarchy process. IEEE Trans. Fuzzy Syst. **22**(4), 749–761 (2014)
59. P. Van Laarhoven, W. Pedrycz, A fuzzy extension of Saaty's priority theory. Fuzzy Sets Syst. **11**(1–3), 229–241 (1983)
60. J.J. Buckley, Fuzzy hierarchical analysis. Fuzzy Sets Syst. **17**(3), 233–247 (1985)
61. D.-Y. Chang, Applications of the extent analysis method on fuzzy AHP. Eur. J. Oper. Res. **95**(3), 649–655 (1996)
62. L. Mikhailov, Deriving priorities from fuzzy pairwise comparison judgements. Fuzzy Sets Syst. **134**(3), 365–385 (2003)
63. J. Rezaei, R. Ortt, V. Scholten, An improved fuzzy preference programming to evaluate entrepreneurship orientation. Appl. Soft Comput. **13**(5), 2749–2758 (2013)
64. Y.-M. Wang, K.-S. Chin, Fuzzy analytic hierarchy process: A logarithmic fuzzy preference programming methodology. Int. J. Approximate Reasoning **52**(4), 541–553 (2011)
65. T.-Y. Hsieh, S.-T. Lu, G.-H. Tzeng, Fuzzy MCDM approach for planning and design tenders selection in public office buildings. Int. J. Project Manage. **22**(7), 573–584 (2004)
66. O. Gogus, T.O. Boucher, Strong transitivity, rationality and weak monotonicity in fuzzy pairwise comparisons. Fuzzy Sets Syst. **94**(1), 133–144 (1998)
67. T.L. Saaty, Absolute and relative measurement with the AHP. The most livable cities in the United States. Socio-Econ. Planning Sci. **20**(6), 327–331 (1986)
68. M.J. Liberatore, R.L. Nydick, P.M. Sanchez, The evaluation of research papers (or how to get an academic committee to agree on something). Interfaces **22**(2), 92–100 (1992)
69. M. Gul, A review of occupational health and safety risk assessment approaches based on multi-criteria decision-making methods and their fuzzy versions. Hum. Ecol. Risk Assess. Int. J. **24**(7), 1723–1760 (2018)
70. A. Khatabakhsh, Z. Maleki, H. Hejazi, M. Pouyakian, Analysis of hazard identification methods in process industries using analytic network process technique (ANP). Iran Occupational Health. **16**(2), 48–60 (2019)
71. F. Omidvari, M. Jahangiri, R. Mehryar, M. Alimohammadlou, M. Kamalinia, Fire risk assessment in healthcare settings: application of FMEA combined with multi-criteria decision making methods. Math. Problems Eng., **2020** (2020)
72. H. Kaleh, M. Pouyakian, A. Abouhossein, Developing a safety rating algorithm for office buildings using Fuzzy Analytical Hierarchy Process (FAHP). Iran Occupational Health J. **18**(1), 313–329 (2021)
73. N. Saffarinia, M. Pouyakian, R. Zendehdel, R. Ramezanian, Presenting a new method to evaluate the severity of the incompatibility of dangerous goods based on FTOPSIS: a case study of the 4.3 class. ACS Chem. Health Safety (2021)

74. A. Rostamabadi, M. Jahangiri, E. Zarei, M. Kamalinia, S. Banaee, M.R. Samaei, A novel fuzzy bayesian network-HFACS (FBN-HFACS) model for analyzing human and organization factors (HOFs) in process accidents. Process Saf. Environ. Prot. **132**, 59–72 (2019)
75. Y. Fu, Y. Qin, L. Kou, X. Liu, L. Jia, Operational risk assessment of railway train based on type-2 intuitionistic fuzzy set and dynamic VIKOR approach. J. Transp. Safety Security, 1–22 (2019)
76. M. Gul, M.F. Ak, A.F. Guneri, Pythagorean fuzzy VIKOR-based approach for safety risk assessment in mine industry. J. Safety Res. **69**, 135–153 (2019)
77. M. Yazdi, A perceptual computing–based method to prioritize intervention actions in the probabilistic risk assessment techniques. Qual. Reliab. Eng. Int. **36**(1), 187–213 (2020)
78. M. Celik, S. Cebi, Analytical HFACS for investigating human errors in shipping accidents. Accid. Anal. Prev. **41**(1), 66–75 (2009)
79. C. La Fata, A. Giallanza, R. Micale, G. La Scalia, Ranking of occupational health and safety risks by a multi-criteria perspective: inclusion of human factors and application of VIKOR. Safety Sci. **138**, 105234 (2021)
80. G.F. Kinney, Practical risk analysis for safety management (1976)
81. M. Sheahan, L. Manion,S. Helper, A Union Guide to QS9000. Detroit: Labor-Management Council for Economic Renewal (1996)
82. S. Standart, Potential Failure Mode and Effects Analysis in Design (Design FMEA), Potential Failure Mode and Effects Analysis in Manufacturing and Assembly Processes (Process FMEA). SAE Standard. Surface Vehicle Recommended Practice SAE-J1739 (2009)
83. M. D. Viner, *Occupational Risk Control: Predicting and Preventing the Unwanted*, Ashgate Publishing, Ltd., (2015)
84. M. Omidvar, F. Nirumand, Risk assessment using FMEA method and on the basis of MCDM, fuzzy logic and grey theory: a case study of overhead cranes. J. Health Safety Work. **7**(1), 63–76 (2017)
85. DIN E, 9241–11-Ergonomics of human-system interaction-Part 11: Usability: Definitions and concepts. (Berlin, 2018)
86. G. Dummer, R. Winton, M. Tooley, An Elementary Guide to Reliability (Elsevier, 1997)
87. A.I. Glendon, S. Clarke, E. McKenna, Human safety and risk management: Crc Press (2016)
88. D. Hoyle, QS-9000 quality systems handbook, Avon Books (1997)
89. ISO I. 2394: 2015–General principles on reliability of structures. International Standard. (2015)
90. I. Mohammadfam, H. Nikoomaram, RETRACTED: FTA vs. Tripod-Beta, which seems better for the analysis of major accidents in process industries? J. Loss Prevent. Process Industries **26**(1), 52–58 (2013)
91. R. E. Smith, MIL-STD-882E. Department of Defence (2012)
92. M. Pouyakian, Safety analysis of glucose production unit in a company by ETBA method and give approaches to prevention of accidents. Tarbiat Modares University (2004)
93. N.A. Stout, D.N. Castillo, H.E. Amandus, T.J. Pizatella, Occupational injuries and work place violence, in *International Encyclopedia of Public Health*, 2nd edn., ed. by S.R. Quah (Academic Press, Oxford, 2017), pp. 334–341
94. Y. Xiang, C. Liu, C. Chao, H. Liu, Risk analysis and assessment of public safety of Submerged Floating Tunnel. Procedia Eng. **4**, 117–125 (2010)

Chapter 5
Fuzzy Sets Theory and Human Reliability: Review, Applications, and Contributions

Kamran Gholamizadeh, Esmaeil Zarei, Mohsen Omidvar, and Mohammad Yazdi

5.1 Introduction

Accident investigations have revealed human error as the most common, unobtrusive, and complicated contributing factor that leads to catastrophic incidents in critical industries such as oil and gas, aviation, nuclear power, transportation [1]. It is reported that more than 80% of accidents in the chemical industry, 75–96% in the marine sector, and over 90% in nuclear power plants have been caused by human error [2]. Furthermore, human factors have a leading role in the life cycle (i.e. construction, operation, maintenance, and emergency management) of numerous systems which make human the most susceptible element of sociotechnical systems to experience failure [1]. After exploring a long journey to figure out how can prevent such major accidents, Human Reliability Analysis (HRA) has been proposed as a proactive approach to systematically recognize, model, and quantify human error which is acknowledged as the main cause of accidents. Human reliability (HR) can be considered as the second major component in complex systems along with technical reliability [3]. Moreover, HR is one of the most complex features of human–machine interface systems due to its unpredictable nature and substantial role in the sustainable

K. Gholamizadeh
Center of Excellence for Occupational Health and Safety Engineering and Research Center of Health Sciences, Hamadan University of Medical Sciences, Hamadan, Iran

E. Zarei (✉)
Centre for Risk, Integrity and Safety Engineering (C-RISE), Faculty of Engineering and Applied Science, Memorial University of Newfoundland, St. John's, NL A1B 3X5, Canada

M. Omidvar
Department of Occupational Health and Safety Engineering, Faculty of Health and Nutrition, Bushehr University of Medical Sciences, Bushehr, Iran

M. Yazdi
Faculty of Engineering and Applied Science, Memorial University of Newfoundland, St. John's, NL A1B 3X5, Canada

M. Yazdi, *Linguistic Methods Under Fuzzy Information in System Safety and Reliability Analysis*, Studies in Fuzziness and Soft Computing 414,
https://doi.org/10.1007/978-3-030-93352-4_5

operation of today's complex systems [4, 5]. As a result, practitioners have actively made their efforts to enhance safety and risk engineering through this perspective since the 1970s. In a simple mathematical definition, human reliability can be defined as Eq. 5.1 [6].

$$HR = \frac{N_e}{N_{eo}} \tag{5.1}$$

where, N_e represents the number of human errors and N_{eo} is the number of error opportunities. As can be seen, HRA is a practical managerial and technical pathway to improve the performance of the system and reduce the occurrence of accidents [7]. It generally is an important subset of probabilistic risk assessment that considers all possible scenarios in system safety. HRA has been often conducted in five steps, which are as follows: (a) definition of the problem or scope of analysis, (b) work or task modeling, (c) human error analysis, (d) human error quantification, and finally, (e) error preventing and mitigating measures, respectively [8].

Different methods are being used and modified to address the drawbacks of previous HRA techniques; they also capture the new aspects of human behavior modeling over recent decades. As examples, absolute probability judgment (APJ), human error assessment and reduction technique (HEART), success likelihood index method (SLIM), cognitive reliability and error analysis method (CREAM), a technique for human event analysis (ATHEANA), cognitive environmental simulation (CES), Bayesian Network (BN)-CREAM, BN-SLIM, which have their own strengths and weaknesses. Conventional methods, despite their wide application, usually have a certain degree of uncertainty which can be attributed to the nature of human error because it is not completely predictable. Hence, there is considerable uncertainty about the quantitative analysis when conventional methods are used on their own [9, 10]. Accordingly, safety experts are looking for tools, approaches, and models that can deal with this uncertainty as much as possible [11]. In this regard, many studies have been done via different tools using probabilistic reasoning models such as Bayesian networks [12–15] and mathematical models such as fuzzy sets [16–20]. These tools have improved the accuracy of HRA by considering nonlinear relationships and complicated interactions between performance shaping factors (PSFs) as well as addressing the data (e.g. human error occurrence, PSFs influence) scarcity in this domain [21, 22].

Fuzzy sets theory is known as one of the most popular tools due to its flexible nature and rigorous mathematical capability to solve the quantification concerns [23]. Its main feature is a great ability to model ambiguous information, which is usually used in modeling systems but is difficult to define precisely [24]. As one of the main difficulties in HRA is the inability to properly define ambiguous parameters such as PSFs, fuzzy sets theory can define and evaluate these parameters in a more reasonable view [25]. Fuzzy logic illustrates well the intrinsic nature of most human communication because it can be considered one of the most similar systems to the human behavioral system [26]. For instance, the human brain in the face of a risk factor can encode several different types of reactions and these can be characterized by linguistic terms in fuzzy sets. Hence, fuzzy sets can be powerful tools in HRA studies because of their similarity to the human brain in decoding linguistic variables. It has been nearly three decades from when fuzzy sets were first applied to HRA, which means it is time to review the conducted literature to build upon it within this research field. Accordingly, the main purpose of this chapter is to review the conducted studies, clarify the application of fuzzy sets in HRA, as well as to find the gaps, challenges, and opportunities for further growth.

5.2 Human Reliability Analysis Review

The HRA was roughly introduced in the 1960s [27] but common methods have been proposed since the mid-1980s. HRA techniques are generally divided into three generations as follows [28]:

First Generation HRA: Given this generation, HRA was performed using only the quantification of Human Error Probabilities (HEPs) and no attention was paid to the in-depth analysis of human behavior especially in cognitive modeling. These techniques include absolute probability judgment (APJ), human error assessment and reduction technique (HEART), justified human error data information (JHEDI), probabilistic human reliability analysis (PHRA), action system operator tree (OATS), and the success likelihood index method (SLIM). The main features of them are the ability to binary represent human actions (success or failure), the phenomenology of human action, emphasis on minimizing the possibility of human wrongdoing, the dichotomy between omission & commission errors, indirect context treatment, and lack of attention to human cognitive actions [29].

Second generation HRA: These methods were developed to overcome the weaknesses of first-generation methods mainly through considering the roots and functions of human cognition with a qualitative description approach. They provided a set of PSFs and how they affect human performance, their qualitative classification, and cognitive analysis of human error. The leading techniques would be cognitive reliability and error analysis method (CREAM), a technique for human event analysis (ATHEANA), cognitive environmental simulation (CES), connectionism assessment of human reliability (CAHR), and method d'evaluation realisation des missions operateur pour la Surete' (MERMOS: assessment method of operational security missions) [29].

Third generation HRA: The two previous generations of techniques suffer from some significant drawbacks such as, being deterministic and failing to address uncertainty, poor ability to be used together with system safety assessment models, failing to present a causal model of PSF in crew performance modeling, and difficulties in using multiple data source (e.g. expert judgment, simulation and observation and types (e.g. crisp, fuzzy) [30]). Accordingly, these shortcomings led to the emergence of a new generation [31]. Bayesian network (BN) has been proposed as a substantial advancement in the third generation of HRA methods which present more insight into HRA data and BN-HRA techniques such as BN-CREAM, BN-SPARH, BN-SLIM.

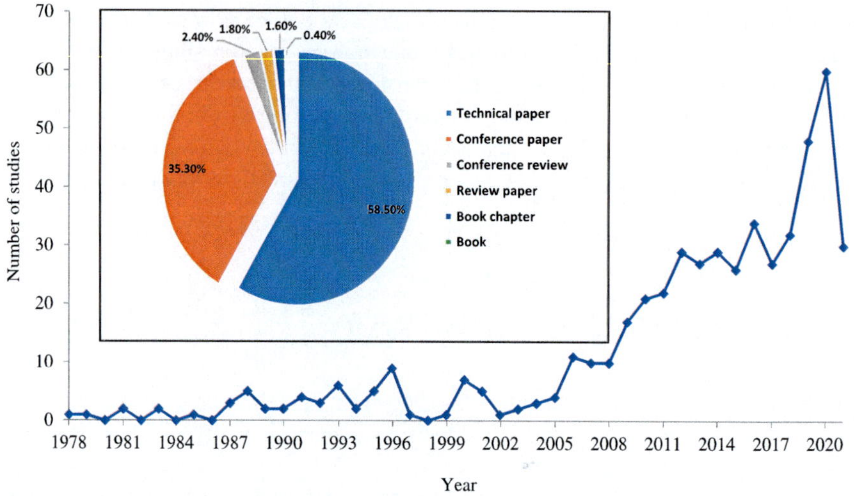

Fig. 5.1 Fuzzy-driven human reliability analysis literature from 1978–202 (Scopus)

It can also analyze human behavior in different time slices [13] using the dynamic Bayesian network (DBN) as the most well-known tool in dynamic evaluation [18].

This chapter will present how fuzzy sets theory improved the common concerns in each aforementioned category. The beginning of fuzzy application in HRA may be associated with Onisawa et al. in 1988 [32] who analyzed the possibility of human error concerning PSFs (e.g. using the fuzzy sets of Zadeh to representing human reliability considering PSFs). Meanwhile, they also conducted their second study [33] examining the role of human error in the Chernobyl accident using fuzzy sets. A bibliometric survey in the Scopus database shows that around 504 studies have been conducted in this field from 1978–2021. Figure 5.1 illustrates the trend of these studies over this period. As can be seen, the upward trend started in 2007 and continued to present. In addition, among these studies, journal articles account for the most (58.2%, 295 studies), while conference papers are in second place (35.2%, 178 studies) (see Fig. 5.1).

5.3 Fuzzy Set Contributions and Applications in HRA

The application of fuzzy logic in HRA is often limited to set theory. The evolution of the fuzzy sets extensions which was first introduced by Zadeh in 1965 [24] is presented in Fig. 5.2. Ten years later (1975) [34], he introduced the type-2 fuzzy set, which had wider applications than the first type because of its concept of using linguistic variables. The application of fuzzy sets by Onisawa can be considered simultaneously with the introduction of multisets fuzzy and Intuisionistic Fuzzy Sets (IFS). Although some recent fuzzy sets have been employed in recent years, the

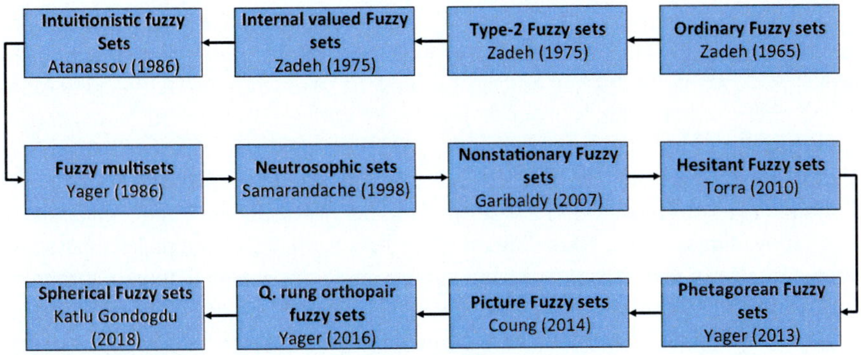

Fig. 5.2 Extensions of fuzzy sets from 1986 to 2018 [39]

fuzzy presented by Zadeh [34] can still be seen in many studies performed on HRA using 1st and 2nd generation methods [5, 17, 20, 35–38].

Concerning utilizing the IFSs in HRA, Tayagi has proposed an IFS success index method in HRA of offshore platform muster scenarios (e.g. man overboard and gas release) considering the various performance shaping factors [40]. Zarei et al. [21] have integrated IFSs theory into the Human Factor Analysis and Classification System (HFACS), and Bayesian network to model human and organizational failures which have occurred in a hot tapping operation of a natural gas pipeline. In another study, Aliabadi [41] integrated these sets into HEART to analyze the human error in a furnace start-up operation. In this study, an Intuitionistic Fuzzy Set is used to estimate the assessed proportion of effect (APOA) in HEART and deal with vagueness and ambiguity in experts' knowledge acquisition. The advantage of the IFSs over ordinary fuzzy sets is that the intuitionistic fuzzy set uses Euclidean geometric functions to determine the distance between non-membership and membership related to the linguistic variables. These calculations are effective in reducing membership uncertainty regarding the language variables [40]. However, despite the application of Neutrosophic fuzzy in technical safety aspects [42–44], it has not been used in HRA yet.

Zhang et al. [45] investigated human error in the nuclear industry using the hesitant fuzzy sets to handle the linguistic assessments given by experts according to the linguistic hesitant fuzzy sets and determine the weights of influential factors by an extended best–worst method, as well as confirming the degree of dependence between successive actions based on the THERP method. The hesitant fuzzy sets are used to express the hesitancy, uncertainties, and inconsistency of experts' judgments. These sets are employed to compute the collective dependency assessment matrix between Human Failure Events (HFEs). Thus, these sets can be used individually or in combination with other fuzzy approaches in studies [45]. In addition, this method can also be used to evaluate the reliability of systems [46, 47]. Furthermore, Pythagorean fuzzy sets have been used in some safety aspects such as risk assessment

[48] and failure analysis [49, 50] and a Pythagorean fuzzy DEMATEL (decision-making trial and evaluation laboratory) model for allocating corrective actions in probabilistic safety analysis in process systems [51]. However, no study on HRA has been employed that. This is also true for Q rung Orthopair fuzzy sets and spherical fuzzy sets.

Spherical fuzzy sets have been recently proposed by Gündoğdu [52] and Kahraman [53] to relax the drawbacks of the previous sets, although it has not to be employed in safety and risk domain. They believe that decision makers can independently define their hesitancies with a larger three-dimensional domain through a linguistic evaluation scale. Therefore, spherical fuzzy sets are one of the most important and underlying extensions to accommodate more uncertainties than existing fuzzy set structures. In these sets, while the squared sum of three parameters (e.g. membership, non-membership, and hesitancy) is between 0 and 1, each of them can be independently estimated from 0 and 1. The superiority comes from collecting the scientifically accepted advantages of other fuzzy sets extensions (e.g. Pythagorean and Neutrosophic sets in a unique theory), while excluding the criticized aspect of Neutrosophic theory, i.e. a sum of membership, non-membership, and hesitancy larger than 1 and the criticized aspect of the Pythagorean theory, i.e. disregarding an independent hesitancy [53, 54]. As a result, these novel fuzzy sets that claim to provide more precise computations regarding knowledge acquisition especially from uncertain data sources, and better capture the potential uncertainty will rapidly rise in popularity.

In the rest of the chapter, the contribution of the aforementioned sets is explained based on the top 6 research interests in HRA which are (1) predicting human error probability, (2) quantifying the influence of performance shaping factors in human error, (3) intra-dependency modeling among these factors, (4) incorporating human error into probabilistic safety and risk analysis, (5) human behavior modeling and finally, (6) characterizing the uncertainty analysis in HRA using fuzzy set theory.

5.3.1 *Predicting Human Error Probability (HEP)*

HEP prediction in a real operation can be considered as the first step to manage human error. It provides a picture of possible errors modes and failure mechanisms. The HEP estimation in complex systems is the most pivotal element of human reliability, which has drawn substantial attention in comparison to other aspects of human behavior and cognitions. Accordingly, considering its importance and the fact that the next measures of human reliability must rely on HEP, it can be seen as the most popular application of fuzzy set theory in HRA. In this regard, numerous studies have been conducted in different industrial sectors as illustrated in Table 5.1. This Table provides useful information concerning the application and contribution of fuzzy sets in predicting human error in various workplaces.

Table 5.1 The applications and contributions of fuzzy sets in predicting HEP in the various sectors

Application		Contributions	Reference(s)
Coal mine		Developing the causal maps based on the fuzzy Z-numbers to predict human error rates in complex systems Addressing the inherent uncertainty in expert knowledge an acquisition Quantifying precisely human error by combining content description factors via the fuzzy mathematical concept	Gupta et al. [55], Kumar et al. [56]
Maritime transportation	Shipping	Applying the fuzzy type-2 to quantify probabilities in decision making	Akyuz et al. [57]
	Oil tankers in ship collision	Quantifying the probabilities of the basic events in the Bayesian network via fuzzy type 2	[58]
	Maritime transportation	Dealing with subjective effects in summarizing the experts' opinions	Erdem et al. 2021) [59]
Land transportation	Offloading operation in oil terminals	Dealing with subjective effects when researchers are faced with a lack of information using fuzzy inference system	Ung et al. [58]
	Oil tanker grounding	Quantifying the probabilities of the basic events via IFSs in the Bayesian network	Gertman et al. [60]
	Hydrocarbon road tanker loading operation	Applying the fuzzy type-2 in handling uncertainties regarding common performance conditions (CPCs)	Ghasemi et al. [61]
Overhead crane operation		Applying the fuzzy sets in summarize the experts' opinions and create a quantitative output from them	Odu et al. [62]
Organizational analysis		Applying the fuzzy sets in analysis the organizational human factors due to the multi-input nature of them and the ability to take quantitative outputs from them	Nemet et al. [63]

(continued)

Table 5.1 (continued)

Application	Contributions	Reference(s)
Airline maintenance	Applying the fuzzy TOPSIS to reduce uncertainty in calculations due to the ambiguity of the concept of human errors in maintenance	Chiu et al. [64]
Nuclear industry site operators	Applying the fuzzy sets type-2 as a computational tool in the absence of information	Ung et al. [65]
Refueling operation in the LPG stations	Applying the fuzzy type-2 to quantify estimation of the experts' opinions	Kumar et al. [23]
Operator error in high-dose-rate treatment	Applying the fuzzy type-2 to quantify estimation of the experts' opinions	Castiglia et al. [66]
Human error in explosive atmosphere (ATEX)	Applying the fuzzy inference system to quantify human errors affected by multiple common performance conditions (CPCs)	Geng et al. [67]
Emergency response in process industries	Applying the fuzzy type-2 to combine CPCs in quantifying human errors probability	Abbasinia et al. [68]
Construction projects	Applying the fuzzy logic rule based (type 2) to estimate the probability of human error under the influence of "accident factors" to reduce uncertainty	Sinabariba et al. [69]
Share systems control	Applying the fuzzy type-2 to describe uncertain cognitive functions	Zhang et al. [70]
Training in nonnuclear industries	Applying the fuzzy type-2 to quantify training parameters which are linguistic in nature	Swaanika et al. [71]
Furnace start-up operation	Applying the IFSs to provide more fuzzy information than ordinary fuzzy sets	Aliabadi et al. [41]

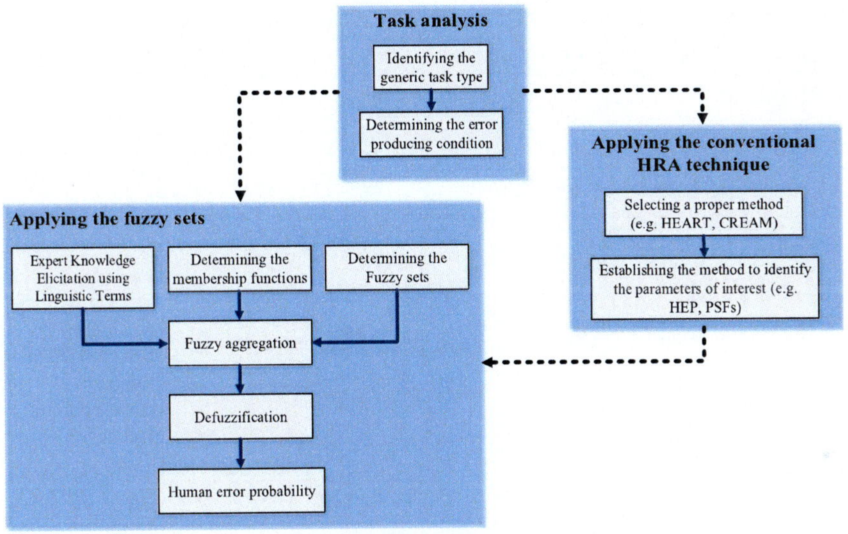

Fig. 5.3 The general structure of fuzzy human error probability analysis

The fuzzy HEP studies have generally followed a similar structure as illustrated in Fig. 5.3. Accordingly, first, the information needed to conduct the study such as the generic task types (GTTs) and error-producing conditions (EPCs) are collected by analyzing the desired tasks. Then, fuzzy set computations based on the framework of the selected HRA technique (i.e. HEART, CREAM, THERP, SPAR-H, etc.) are used for quantifying the parameters of interest such as expert weight, the importance level of contributing factors, and then HEP.

Table 5.2 presents the mathematical expression to estimate HEP in some common HRA techniques. The associated parameters, which are the most interesting to use fuzzy set theory for establishing a rigorous computation, in each technique are shown in the last column.

The HEART, as one of the most popular HRA methods, is selected to illustrate the application of fuzzy theory in predicting HEP. The mathematical expression for estimating HEP in this method is presented by Eqs. 5.2 and 5.3.

$$HEP = GEP_j \times \prod_{1}^{38} WF_i \tag{5.2}$$

$$WF_i = APOA_i \times (EPC_i - 1) \quad for\ i = 1, 2, 3, \ldots, n;\ n \leq 38 \tag{5.3}$$

Table 5.2 Mathematical expression to estimate HEP in some common HRA techniques

Method	Mathematical expression to estimate HEP	Parameters
ATHENA	$P(EIS) = \sum_{UA_i} \sum_{UFC_j} P_{ij}(S)$	UA_i: unsafe acts UFC_j: unsafe forcing context P_{ij}: unsafe act probability
SPAR-H	$HEP = \frac{NHEP.\prod_1^8 S_i}{NHEP.\left(\prod_1^8 S_i - 1\right) + 1}$	NHEP: nominal human error probability S_i: multiplier associated with the score of corresponding PSF levels
HEART	AHE = ((Multiplier of EPC − 1) × Assessed proportion of the effect) + 1 HEP = nominal human unreliability · $\prod$ AHE_i	AHE: Assessed Human Effect EPC: Error-Producing Conditions
THERP	$HEP_{Final} = BHEP_{Diagnosis} \cdot \prod_1^{\pi} PSF_{Diagnosis.i} + BHEP_{Execution} \cdot \prod_1^{\pi} PSF_{Execution.i}$	PSFs' Performance Shaping Factors BHEP: Basic Human Error Probability

where WF_i = weight factor for the ith *EPC*; EPC_i = EPC for the -th condition; $APOA_i$ = *APOA* for -th condition; *HEP* = Final *HEP*; GEP_j = *GEP* (the basic error likelihood which is attributed to each *GTT*) connected with to be evaluated task. In this regard, Table 5.3 shows the description of generic task type and related GEP value. Moreover, Table 5.4 shows the EPCs definition and related maximum value based on the HEART framework.

Among the required parameters to predict HEP, APOA is often estimated by subject matter experts' judgments which may be subjected to ambiguity and vagueness. An intuitionistic Fuzzy (IF) set is chosen to deal with this issue during a furnace start-up operation [41] where numerous major fires and explosions are caused due to human error. To this end, experts are first asked to express their belief regarding the parameter of interest (e.g., APOA). For this purpose, first, the tasks and sub-tasks related to furnace start-up operation should be determined using hierarchical task analysis. Table 5.5 shows these tasks.

After collecting the primary data, fuzzy computations are performed as follows:

Step 1: Estimating expert weighting using Fuzzy- Analytical Hierarchy Process (FAHP).

Table 5.3 Description of generic task type and related GEP value [72]

GGT ID	Generic Task Type (GTT)	GEP
A	Totally unfamiliar, performed at speed with no real idea of the likely consequences	0.55
B	Shift or restore system to a new or original state on a single attempt without supervision or procedures	0.26
C	complex task requiring a high level of comprehension and skill	0.16
D	Fairly simple task performed rapidly or given scant attention	0.09
E	Outline, highly practiced, and rapid task involving a relatively low skill level	0.02
F	Restore or shift the system to an original or new state following procedures, with some checking	0.003
G	Completely familiar, well-designed, highly practiced, and routine task occurring several times per hour, performed to the highest possible standards by highly motivated, highly trained, and experienced persons, totally aware of the implications of failure and having the time to correct potential errors but without the benefits provided by significant job aids	0.0004
H	Respond correctly to system commands even when there is an augmented or automated supervisory system providing an accurate interpretation of the system state	0.00002
M	Miscellaneous task for which no description can be found	0.03

Let us assume that 5 experts were consulted in the fuzzy calculations.

1. After constructing the pairwise comparison matrix, the fuzzy weights are calculated by Eq. 5.4.

$$\widetilde{a_i} = (\widetilde{a_{i1}} \times \widetilde{a_{i2}} \times \ldots \times \widetilde{a_{in}})^{1/n} \quad (5.4)$$

For instance, for $\widetilde{a_{12}}$ is calculated as:

$$\begin{aligned}\widetilde{a_{12}} &= \left((1, 1, 3) \times (1, 3, 5) \times \left(\frac{1}{5}, \frac{1}{3}, 1\right) \times (1, 3, 5)\right)^{1/4} \\ &= \left(\left(1 \times 1 \times \frac{1}{5} \times 1\right)^{1/4}, \left(1 \times 3 \times \frac{1}{3} \times 3\right)^{1/4}, (3 \times 5 \times 1 \times 5)^{1/4}\right) \\ &= (0.668, 1.31, 2.94)\end{aligned}$$

After all the $\widetilde{a_{ij}}$ values are calculated in the same way, the synthetic pairwise comparison matrix of the expert abilities was made.

2. Compute the fuzzy weights of each dimension as Eq. 5.5.

Table 5.4 EPC definition and related maximum value [41]

EPC ID	Error producing condition	Maximum amount
EPC1	Unfamiliarity with a situation that is potentially important but which only occurs infrequently or which is novel	× 17
EPC2	A shortage of time available for error detection and correction	× 11
EPC3	A low signal–noise ratio	× 10
EPC4	A means of suppressing or over-riding information or features that is too easily Accessible	× 9
EPC5	No means of conveying spatial and functional information to operators in a form which they can readily assimilate	× 8
EPC6	A mismatch between an operator's model of the world and that imagined by a Designer	× 8
EPC7	No obvious means of reversing an unintended action	× 8
EPC8	A channel capacity overload, particularly one caused by simultaneous presentation of no redundant information	× 6
EPC9	A need to unlearn a technique and apply one which requires the application of an opposing philosophy	× 6
EPC10	The need to transfer specific knowledge from task to task without loss	× 5.5
EPC11	Ambiguity in the required performance standards	× 5
EPC12	A mismatch between perceived and real risk	× 4
EPC13	Poor, ambiguous or ill-matched system feedback	× 4
EPC14	No clear direct and timely confirmation of an intended action from the portion of the system over which control is to be exerted	× 4
EPC15	Operator inexperience (e.g., a newly qualified tradesman, but not an "expert")	× 3
EPC16	An impoverished quality of information conveyed by procedures and person/person interaction	× 3
EPC17	Little or no independent checking or testing of output	× 3
EPC18	A conflict between immediate and long-term objectives	× 2.5
EPC19	No diversity of information input for veracity checks	× 2.5
EPC20	A mismatch between the educational achievement level of an individual and the requirements of the task	× 2
EPC21	An incentive to use other more dangerous procedures	× 2

(continued)

Table 5.4 (continued)

EPC ID	Error producing condition	Maximum amount
EPC22	Little opportunity to exercise mind and body outside the immediate confines of a Job	× 1.8
EPC23	Unreliable instrumentation (enough that it is noticed)	× 1.6
EPC24	A need for absolute judgments which are beyond the capabilities or experience of an operator	× 1.6
EPC25	Unclear allocation of function and responsibility	× 1.6
EPC26	No obvious way to keep track of progress during an activity	× 1.4
EPC27	A danger that finite physical capabilities will be exceeded	× 1.4
EPC28	Little or no intrinsic meaning in a task	× 1.4
EPC29	High-level emotional stress	× 1.3
EPC30	Evidence of ill-health amongst operatives, especially fever	× 1.2
EPC31	Low workforce morale	× 1.2
EPC32	Inconsistency of meaning of displays and procedures	× 1.2
EPC33	A poor or hostile environment (below 75% of health or life-threatening severity)	× 1.5
EPC34	Prolonged inactivity or highly repetitious cycling of low mental workload tasks (For 1st half hour)	× 1.1
EPC35	(Thereafter)	× 1.05
EPC36	Disruption of normal work-sleep cycles	× 1.1
EPC37	Task pacing caused by the intervention of others	× 1.06
EPC38	Additional team members over and above those necessary to perform task normally and satisfactorily (per additional man)	× 1.03

$$\widetilde{r_i} = (\widetilde{a_{i1}} \times \widetilde{a_{i2}} \times \ldots \times \widetilde{a_{in}})^{1/n}(5.5) \qquad (5.5)$$

Based on it, the weights of the criteria for five experts are found. For example, the dimension of expert#1$(\widetilde{r_1})$ was calculated as:

$$\begin{aligned}\widetilde{r_1} &= (\widetilde{a_{11}} \times \widetilde{a_{12}} \times \widetilde{a_{13}} \times \widetilde{a_{14}} \times \widetilde{a_{15}})^{1/5}\\ &= \big((1 \times 0.668 \times 1.96 \times 2.43 \times 6.14)^{1/5}, (1 \times 21.31 \times 4.21 \times 4.87 \times 0.880)^{1/5},\\ &\quad (1 \times 2.94 \times 6.29 \times 6.70 \times 1.96)^{1/5}\big)\\ &= (1.14, 1.48, 2.45)\end{aligned}$$

All the $\widetilde{r_i}$ values were obtained as Table 5.6.

3. Compute the fuzzy weights of each criterion as Eq. 5.6

Table 5.5 Human task analysis for the furnace start-up operation [41]

Subtask ID	Task description
1	*Pre-Start up Safety Review*
A1	Make sure that pre-startup checklists are completed
2	*Preparation*
A2	Make sure furnace been boxed up
A3	Check if fuel-system valves are in tightly closed position
A4	Full open secondary air damper of burners and stack damper
A5	Turn on the forced draft fan and make sure firebox full purged
A6	Check if all blank in pilot, fuel gas and waste gas systems are in position
A7	Purge fuel line to each burner with nitrogen
A8	Pressurize each fuel line with nitrogen and check for leakage
A9	Blank off the pilot fuel gas lines
A10	Drain the liquid from the main fuel gas knockout drum
A11	Keep a safe minimum flow through the furnace tubes
3	*Pilots' ignition*
A12	Check stack draft
A13	Care must be taken to minimize the number of people around the furnace
A14	Make sure that there is no flammable gas in the fire chamber
A15	Push igniter of the first pilot and lit remaining the way you lit the first one
4	*Burner's ignition*
A16	Check if all pilots are lighted
A17	Blank off the main fuel gas and waste gas line
A18	Recheck the knockout drum and drain all liquid
A19	Recheck stack draft
A20	Shut off the primary damper of each burner before lighting
A21	Open the gas control valve slowly just before lighting the first burner
A22	Make sure that lighting of remaining burners is even spread
A23	Make sure that furnace reach normal operating conditions before lighting any waste gas burner

Table 5.6 Calculated $\widetilde{r_i}$ values for each expert

$\widetilde{r_1}$	$\widetilde{r_2}$	$\widetilde{r_3}$	$\widetilde{r_4}$	$\widetilde{r_5}$
(1.14,1.48.2.45)	(1.03,1.71,2.74)	(0.403,0.452,0.973)	(0.317,0.356,0.774)	(1.04,1.71,2.70)

$$\widetilde{w_i} = \widetilde{r_i}(\widetilde{r_1} + \widetilde{r_2} + \ldots + \widetilde{r_n})^{-1} \tag{5.6}$$

The weights of the all criterion were calculated. For example, the criterion of expert#1 ($\widetilde{w_1}$) was calculated as:

Table 5.7 Calculated $\widetilde{w}_i$ values for each expert

$\widetilde{w}_1$	$\widetilde{w}_2$	$\widetilde{w}_3$	$\widetilde{w}_4$	$\widetilde{w}_5$
(0.254, 0.159, 0.293)	(0.262, 0.271, 0.268)	(0.102, 0.071, 0.095)	(0.080, 0.056, 0.075)	(0.264, 0.271, 0.264)

$$\begin{aligned}\widetilde{w}_1 &= \overline{r}_i(\widetilde{r}_1 + \widetilde{r}_2 + \widetilde{r}_3 + \widetilde{r}_4 + \widetilde{r}_5)^{-1}\\ &= (1.14, 1.48, 2.45) \times \frac{1}{1.14 + 1.03 + 0.403, 0.317, 1.04}\\ &\times \frac{1}{1.88 + 1.71 + 0.452 + 0.356 + 1.71}\\ &\times \frac{1}{3 + 2.74 + 0.973 + 0.774 + 2.7} = (0.254, 0.159, 0.293)\end{aligned}$$

All the $\widetilde{w}_i$ values were obtained as Table 5.7:

4. Finally, the weights of each expert are obtained by the center of gravity defuzzification method of $\widetilde{a}_i$ using Eq. 5.7

$$Defuzzified(w_i') = \frac{[(uw_i - lw_i) + (mw_i - lw_i)]}{3} + lw_i \tag{5.7}$$

In the end, the weight of each expert was calculated using the defuzzification process. For example, w_1' was calculated as:

$$w_1' = \frac{[(0.293 - 0.159) + (0.254 - 0.159)]}{3} + 0.159 = 0.388$$

All the weights of experts were obtained as Table 5.8.

5. Normalize the calculated crisp weights as Eq. 5.8.

$$w_i = \frac{w_i'}{\sum_{i=1}^{n} w_i',\ i = 1, \ldots, n} \tag{5.8}$$

Since the summation of all expert weight (1.082), are greater than one, the normalized weight of expert weights was found as follows:

$$w_{ej} = (0.358, 0.246, 0.082, 0.064, 0.0250)$$

Step 2: Expert elicitation regarding APOA.

Table 5.8 Calculated weight for each expert

$\tilde{w}_1$	$\tilde{w}_2$	$\tilde{w}_3$	$\tilde{w}_4$	$\tilde{w}_5$
0.388	0.267	0.089	0.070	0.268

Table 5.9 Linguistic terms and their corresponding IFN to assess the relative strength of EPCs (APOA) [41]

Linguistic term	Conversion scale
Very low (VL)	(0.00, 0.04, 0.08; 0.00, 0.04, 0.08)
Low (L)	(0.07, 0.13, 0.19; 0.06, 0.13, 0.2)
Mildly low (ML)	(0.17, 0.27, 0.37; 0.15, 0.27, 0.39)
Medium (M)	(0.35, 0.50, 0.65; 0.32, 0.50, 0.68)
Mildly high (MH)	(0.62, 0.73, 0.82; 0.61, 0.73, 0.85)
High (H)	(0.81, 0.87, 0.93; 0.79, 0.87, 0.95)
Very high (VH)	(0.92, 0.96, 1,00; 0.92, 0.96, 1.00)

Table 5.10 The opinion of each expert about the A.20-EPC4

Experts' opinions about A.20-EPC4	
Expert 1: (ML)	(0.17, 0.27, 0.37; 0.15, 0.27, 0.39)
Expert 2: (ML)	(0.17, 0.27, 0.37; 0.15, 0.27, 0.39)
Expert 3: (VL)	(0.00, 0.04, 0.08; 0.00, 0.04, 0.08)
Expert 4: (VL)	(0.00, 0.04, 0.08; 0.00, 0.04, 0.08)
Expert 5: (L)	(0.07, 0.13, 0.19; 0.06, 0.13, 0.2)

Table 5.9 illustrates linguistic terms and their corresponding IFN to assess the relative strength of EPCs (APOA) in the defined scenario.

Step 3: Aggregation: For example, the aggregation process for A20-EPC4 was conducted as:

A20: Shut off the primary damper of each burner before lighting.

EPC4: A means of suppressing or over-riding information or features which is too easily accessible.

Table 5.10 shows the opinion of each expert about the A.20-EPC4.

1. Estimating a degree of similarity between the experts using Eq. 5.9.

$$S\left(\tilde{A}, \tilde{B}\right) = 1 - \frac{1}{J}\sum_{i=1}^{J}|a_i - b_i| \tag{5.9}$$

For example, to estimate a degree of similarity between expert1 and expert2 (S (E1&E2)):

$$S_{E1,E2} = 1 - 1/4(0.58 + 0.58 + 0.58 + 0.58) = 0.420$$

Based on it, the degrees of similarity (S) of all experts are provided in Table 5.11.

2. Estimating the average of agreement (AA) using Eq. 5.10.

Table 5.11 The degrees of similarity (S) of all experts

S(E1&E2)	1.000	S(E2&E4)	0.148
S(E1&E3)	0.148	S(E2&E5)	0.308
S(E1&E4)	0.148	S(E3&E4)	1.000
S(E1&E5)	0.308	S(E3&E5)	0.308
S(E2&E3)	0.148	S(E4&E5)	0.308

$$AA(E_u) = \frac{1}{m-1} \sum_{\substack{u \neq v \\ v = 1}}^{J} s\left(\tilde{R}u, \tilde{R}v\right) \tag{5.10}$$

For instance, to estimate the average of agreement of expert1:

$$1/(5-1)(1.000 + 0.148 + 0.148 + 0.308) = 0.401$$

The averages of agreement of all experts are shown in Table 5.12.

3. Estimating the relative agreement using Eq. 5.11.

$$RA(E_u) = \frac{AA(E_u)}{\sum_{u=1}^{J} AA(E_u)} \tag{5.11}$$

For example, to estimate the relative agreement (RA) of expert1:

$$0.401/(0.401 + 0.401 + 0.688 + 0.688 + 0.308) = 0.1293$$

The relative agreements of all experts are provided in Table 5.13.

4. Estimating the consensus coefficient using Eq. 5.12.

$$CC(E_1) = \beta \cdot W(E_u) + (1-\beta) \cdot RA(\mathrm{E}_1) \tag{5.12}$$

For example, to estimating the consensus coefficient (CC) of expert1:

Table 5.12 The averages of agreement of all experts

AA (E1)	AA (E2)	AA (E3)	AA (E4)	AA (E5)
0.401	0.401	0.688	0.688	0.308

Table 5.13 The relative agreements of all experts

RA (E1)	RA (E2)	RA (E3)	RA (E4)	RA (E5)
0.1293	0.1293	0.2218	0.2218	0.0993

Table 5.14 The consensus coefficients of all experts

CC (E1)	CC (E2)	CC (E3)	CC (E4)	CC (E5)
0.265	0.265	0.455	0.455	0.203

$$0.5 \times 0.401 + 0.5 \times 0.1293 = 0.265$$

The consensus coefficients of all experts are shown in Table 5.14.

5. Aggregation of experts' opinion using Eq. 5.13.

$$\tilde{\mathrm{R}}_{\mathrm{AG}} = \mathrm{CC}(\mathrm{E}_1) \times \tilde{\mathrm{R}}_2 + \mathrm{CC}(\mathrm{E}_2) \times \tilde{\mathrm{R}}_2 + \c{d}ots + \mathrm{CC}(\mathrm{E}_\mathrm{m}) \times \tilde{\mathrm{R}}_\mathrm{M} \tag{5.13}$$

For our case (A.20-EPC4) we have:

$$\begin{aligned} &0.265 \times (0.17, 0.27, 0.37; 0.15, 0.27, 0.39) \\ &\quad + 0.265 \times (0.17, 0.27, 0.37; 0.15, 0.27, 0.39) \\ &\quad + 0.455 \times (0.00, 0.04, 0.08; 0.00, 0.04, 0.08) \\ &\quad + 0.455 \times (0.00, 0.04, 0.08; 0.00, 0.04, 0.08) \\ &\quad + 0.203 \times (0.07, 0.13, 0.19; 0.06, 0.13, 0.2) \\ &\quad\ = (0.104, 0.205, 0.307; 0.091, 0.205, 0.320) \end{aligned}$$

6. Defuzzification:

Defuzzification of IFN AE = (a, b, c; a′, b, c′) can be defined as Eq. 5.14. IFN AE = (a, b, c; a′, b, c′): (0.349, 0.416, 0.483; 0.335, 0.416, 0.498):

$$\begin{aligned} X &= \frac{1}{3}\left(\frac{(\mathrm{c}' - \mathrm{a}')(b - 2\mathrm{c}' - 2\mathrm{a}') + (\mathrm{c} - \mathrm{a})(\mathrm{a} + \mathrm{b} + \mathrm{c}) + 3(\mathrm{c}'^2 - \mathrm{a}'^2)}{\mathrm{c}' - \mathrm{a}' + c - a}\right) \\ &= 0.0310 \end{aligned} \tag{5.14}$$

7. HEP calculation using equation.

Finally, to calculating the probability we use Eqs. 5.15 and 5.16

$$\text{probability of IR} = \begin{cases} 1/10^k & IR \neq 0 \\ 0 & R = 0 \end{cases} \tag{5.15}$$

$$k = 2.301 \times [(1 - IR)/IR]^{1/3} \tag{5.16}$$

In our case, the HEP of A20-EPC4.
$k = ((1/0.0310) - 1)^{1/3} \times 2.301 = 7.248$ and probability $= \frac{1}{10^{7.248}} = 5.65E - 8$.
The above steps can be repeated for all sub-tasks and their related EPCs to estimate the associated HEP. More information concerning this case study is available in [41].

5.3.2 *Quantifying the Performance Shaping Factors (PSFs) Influence*

In general, PSFs that affect the likelihood of human error can significantly vary depending on the context and importance of human activities. There is a direct relationship between the complexity of human tasks and the diversity of PSFs and determining the importance of each of them. The results can be considered as proper guides for safety experts to prevent human error. The estimated HEP has been often modified by taking PSFs influence into account since human performance is substantially affected by various task, machine, environment, organization, and technology-based factors in real circumstances [1]. However, assigning a proper value associated with these factors, which reflects a real impact and importance (weight) of them on operator performance is one of the most interesting and also challenging research subjects [1]. These factors can be categorized as an organization, information (knowledge), job design, human-system interface, work environment, workplace design, and operator characteristics [73]. Table 5.15 presents those factors in fifteen taxonomies related to PSFs. They present a great diversity in terms of quality and quantity of data collection for quantifying to what extent they influence operator performance.

Assigning an optimal weight for the contributing factors is a challenge that arises commonly in most of the multi-criteria decision-making (MCDM) techniques. It is a fact that different influencing factors don't same importance on all real conditions and those values substantially can influence the outcome (e.g. over or under estimation) of the decision-making process. Accordingly, it is vital to pay particular attention to properly weight the factors in estimating HEP. To this end, simulation, observation, historical field data, and subject-matter experts' knowledge elicitation can be a primary sources to collect data, although there are serious practical issues to hinder them in reality. In the conventional HRA techniques, predefined values are assigned as PFSs influence which SPAR-H is illustrated as an example in Table 5.16.

Considering procedure (status = misleading, multiplier = 10), ergonomics (status = poor, multiplier = 20) and complexity (status = moderately complex, multiplier = 2) as three contributing factors on operator error probability in a control room. According to SPAR-H method guideline [60], PSF composite score (see Table 5.16) is 400 ($10 \times 20 \times 2$). Those predefined scores simplify the computation process. However, considering their importance in numerical outcomes and assigning safety measures for damping the influence of critical contributing factors to reduce HEP. This computation can provide misleading results and suffers from serious drawbacks. Some of them can be as (a) Different PSFs doesn't have the same importance in real operation and it should be considered along with its influence, score, multiplier or probability, (b) Conventional HRA guidelines suffer from an incomplete list of PSFs, subsequently their scores and fail to reflect many contributing factors associated with industry 4.0 (e.g. digitization level) and non-behavior factors (e.g. Moral, attitude, safety culture), (c) simple linear mathematical expressions used to aggregate scores of various PSFs and not reflect a rigorous computation. To relax it, fuzzy set theory has

Table 5.15 Some representative PSFs from surveyed PSF taxonomies (Name of PSF taxonomy (Number of total PSFs)) [74]

THERP (67)	**SLIM** (7)	**INTENT** (11)	**HEART** (38)	**CREAM** (9)
Quality of environment	Role of operations	Stress	Emotional stress	Working condition
Work hours	Quality of decision	Experience	Low workforce moral	Available time
Decision making	Moral/motivator	Safety culture	Age of personnel	Circadian rhythm
Task criticality	Teams	Training	Poor environment	Crew collaboration
Sex differences	Stress	Motivation	Unreliable instrument	Training and experience
Physical condition	Competence	workload	Lack of available time	Adequacy of organization
HRMS (6)	ATHEANA (9)	IDAC (53)	K-HRA (15)	CBDT (30)
Time	Procedure	Tool availability	Primary task or not	Workload
Quality of information	Communication	Procedure availability	Level of alarm/HMI	Formal communication
Training experience, familiarity	Supervision	Time constraint load	Level of procedure	Specific training
Degree of adequacy of procedure	Staffing	Awareness of role	Time urgency	Practiced scenario
Task organization	Human-system inference	Problem solving style	Training and education	Reasonable alternative
Task complexity	Organizational factors	Physical limitation	Level of supervision	Training on step
HCR (3)	SPAR-H (39)	INCORECT (8)	Bellami's PSF taxonomy (8)	Gerde's taxonomy (108)
Operator experience	Mental stress	Available time	Individual factors	Communication
Stress level	Available time of act	Availability of plans	Task demands	Decision support
Quality of operator/interface	Years of experience	Simultaneous goals	Instruction and procedure	Fatigue
	Personal problem	Working condition	Stress	Motivation
	Safety culture	Performance feedback	Environment	Hardware reliability
	Communication	Information uncertainty	MMI characteristics	Quality of procedure

Table 5.16 SPAR-H PSFs used in quantifying HEPs [60]

PSF	Diagnosis levels	Action levels	Range of influences
Available time	5	5	0.01 to failure
Stress/stressors	3	3	1 to 5
Complexity	3 (4)[a]	3	0.1 to 5
Experience/training	3	3	0.5 to 3
Procedures	4	4	0.5 to 50
Ergonomics/HMI	4	4	0.5 to 50
Fitness for Duty	3	3	1 to failure
Work processes	3	3	0.8 to 5

[a] The number in parentheses = the number of levels associated with LP/SD

been significantly raised in popularity in recent years. It uses rigorous mathematical computations to quantify subject-matter experts' knowledge, which can be obtained in all real situations, under uncertainty. Two types of values are corresponding with PSFs which can be estimated using fuzzy sets. Firstly, to what extent do these factors influence (impact) on human unreliability (error probability), and then how much should be assumed of their importance (weight) in human performance.

Weighting methods related to MCDM are divided into three categories: subjective weighting methods, objective weighting methods, and integrated weighting methods [62]. The fuzzy approach is used to reduce subjectivity. Therefore, in this section, first, the characteristics of each method are expressed, and then by reviewing the studies, challenges in the present studies are identified and finally, suggestions for future studies are provided. Table 5.17 shows the characteristics of the most common subjective methods for these three categories.

Studies in this field have been conducted in various industrial sectors. Kovacevic et al. [76] quantified the weight of PSFs in maintenance of mining machines using FAHP. They used the opinions of experts in the brainstorming process and then, with consensus, they calculated the weight of each factor using a fuzzy decision matrix (based on crisp value and linguistic variables) and the row geometric mean method. Ung et al. [77] weighed and prioritized PSFs related to the port operation in marine accidents using FAHP based on the Weighted Mean of Maximums defuzzification. They asked the experts to compare the pairs of factors based on a fuzzy guide argument and then calculated the weight of the factors by aggregating the experts' opinions. The main reason for using fuzzy sets in this study was the ability of fuzzy AHP to quantify the expert's opinions in the brainstorming process. Hu et al. [78] weighed PSFs in Shale gas fracturing operation. They first obtained the relevant PSFs through a Hazard and Operability Analysis process and then weighed and prioritized each using a "fuzzy-VICOR (Vlse Kriterijumska Optimizacija Kompromisno Resenje) based on the best or worst solution. They hypothesized that the fuzzy-VICOR is appropriate tool because of its ability to prioritize PSFs when faced with various conflicting variables.

Table 5.17 Characteristics of subjective weighting methods in multi-criteria decision making [62, 63, 75]

Method	Characteristics
Analytical Hierarchy Process (AHP)	One of the most common methods in MCDM techniques Easy to use Proper accuracy despite the ease of use Difficulty recognizing the method of processing the information obtained by the decision maker Limited use when high numbers pairwise comparisons were needed
Elimination and Choice Translating Reality (ELETRE)	Ranking alternatives from best to worst using pairwise comparisons Ability to delete less favorable alternatives It is useful just when the decision maker needs to choose fewer criteria with many alternatives
Vlse Kriterijumska Optimizacija Kompromisno Resenje' (VIKOR)	Developed to solve decision problems with conflicting and incomparable criteria, assuming that compromise is acceptable for conflict resolution Focuses on ranking and selecting from a set of options and determines the compromise solution that is closest to the ideal solution via the best or worst solution
Technique of Ranking Preferences by Similarity to the Ideal Solution (TOPSIS)	It is a method of compensatory aggregation Compares a set of alternatives by identifying weights for each criterion and normalizing scores for each criterion Calculates the geometric distance between each option and the ideal substitute, which is the best score in each criterion Allows exchange between criteria, if a poor result in one criterion with a good result in another is ignored Provides a more realistic form of modeling than non-compensatory methods
Analytic Network Process (ANP)	Generalization of the AHP is by considering the dependencies between the elements of the hierarchy Many decision problems cannot be hierarchically structured because they involve the interaction and dependence of higher-level elements in the hierarchy on lower-level elements. Thus, ANP is represented by a network instead of a hierarchy
Simple Multi-attribute Ranking Technique (SMART)	Describes alternatives and weight criteria In this method, the decision maker is asked to rank the criteria in terms of importance from the worst (score = 10) to the best (score = 100) and the weight of the criteria is calculated by normalizing the total score to 1

(continued)

Table 5.17 (continued)

Method	Characteristics
Best–worst method (BMW)	It is a pairwise comparison method that offers a structured way of comparing Identifying the best and worst criteria (or alternatives) before making even comparisons between criteria (or alternatives) Reducing the possible bias during pairwise comparisons using two pairwise comparison vectors based on two opposing references (best and worst) in a single optimization model It is an efficient method that can also provide the possibility of examining the stability of the presented pairwise comparisons Having multiple optimal solutions and flexibility in a situation where multiple decision-maker(s) are involved

Studies like these were performed in process industries as well. Dos Santos et al. [79] weighed PSFs associated with offshore operations from subsea wells to refineries. By combining AHP and IFSs, they prioritized the factors using the opinions of three experts. The main reason they used this set was because of the ability of intuitionistic FAHP to weigh PSFs when faced with quantitative and qualitative aspects. Moreover, the ability of intuitionistic FAHP to reduce uncertainty in eliciting expert opinions using triangular membership functions was demonstrated by Zarei et al. [21]. They used these sets to estimate the influence of various contributing factors which can lead to an occupational accident during hot tapping operations in gas pipelines. In this study, FST is also integrated onto Bayesian Networks to deal with the parameter and model uncertainty. Another aspect that has drawn the attention of experts is the applicability of IFSs in quantitative evaluation of known sub-tasks in the system and weighting them with low uncertainty in comparison to the conventional methods [40]. In this regard, the ability of IFSs to handle randomness and consistency between the experts can be considered. The main purpose of these mentioned studies was to find PSFs with the greatest impact on human error. The common approach of all the studies was to first the PSFs affecting the identified sub-tasks are determined. Then, the weight and priority of PSFs are determined via experts, opinions. Some methods such as intuitionistic fuzzy, fuzzy Technique of Ranking Preferences by Similarity to the Ideal Solution (fuzzy-TOPSIS), fuzzy AHP, and fuzzy VICOR have been used in the mentioned studies (see Table 5.17), but no qualitative or quantitative comparison has been made between conventional weighting methods. This author suggests that in future studies, the reliability indices of each of them be measured and compared. The highest number of studies in the PSFs weighting field is related to the application of FAHP. The reason for this can be attributed to the ease of use of AHP and its acceptable accuracy in combination with fuzzy sets. Combining fuzzy sets with the best–worst method for the reasons mentioned can be of more interest to researchers in future studies. However, it is also recommended to use the Simple Multi-attribute Ranking Technique method and its extension using fuzzy sets.

5.3.3 Modeling Intra-dependency Among PSFs

An accurate reflection of PSF influence on performance variability requires modeling the interactions among those factors, which is one of the most interesting and newest aspects of HRA studies. However, considering numerous diverse tasks, machines, environments, organizations, and technology-based factors, figuring out how and to what extent there are direct and indirect (latent) integrations among those contributing factors is problematic. Accordingly, dealing with the entire these integrations to first develop a causality map, which is hard to use the conventional HRA techniques. This is also true using the other data collection procedures such as simulation, field data, and observation concerning limited resources (e.g. time, budget, and human). The MCDM methods play a substantial role in addressing this important issue and have risen in popularity in recent years in most academic concerns. Table 5.18 provides the conducted investigations in this matter, although some of them may not use for HRA directly.

As can be seen, fuzzy extensions of the Analytic Network Process (ANP), AHP, and DEMATEL are more for quantitatively developing a casual map of influencing factors. These also take aleatory uncertainty into account in modeling the PSFs' influence and subsequently HRA. Studies have shown that the use of ANP can account for possible interactions between elements that are ignored in the AHP method but are not able to determine the strengths of the relationships between the criteria. Failure to pay attention to this issue causes the model results to deviate from the real conditions. To overcome this shortcoming, DEMATEL in combination with ANP (ANP-DEMATEL) can provide more accurate results. Another reason for the relevance of this combination is that the use of DEMATEL is used to plot relationships between criteria in ANP and can increase the accuracy of determining the relative importance of criteria in evaluations [86, 87]. This advantage has also been confirmed for the ANP-TOPSIS combination, where ANP weighting and TOPSIS prioritization have provided more accurate results [88]. It is suggested that future studies consider combining other weighting methods such as Best–Worst Method, Simple Multi-attribute Ranking Technique, and Elimination and Choice Translating Reality with DEMATEL in a fuzzy framework.

The main application of fuzzy set theory and fuzzy MCDM methods in human error analysis is illustrated in Fig. 5.4. Fuzzy sets are used to estimate HEP, while fuzzy extensions of MCDM techniques are employed to capture influence, importance, and intra-dependency associated with PSFs. Among the weighting methods, it seems that BWM can be more efficient than other methods because BWM using two pairwise comparison vectors based on two opposing references (best and worst) in a single optimization model can reduce the possible bias of decision-maker anchoring during pairwise comparisons. In pairwise comparison methods, we either

Table 5.18 Intra-dependency modeling among PSFs using fuzzy MCDM methods

Application		Contribution	Reference
Steam boiler operation in the process industry		Presenting DEMATEL application to overcome limitations in real-life errors calculation by displaying complex relations between main tasks, subtasks, and error producing conditions (EPCs)	Can et al. [80]
Railway network	High-speed railway dispatching tasks	Handling interdependency of interactions between EPCs and subsequently reducing uncertainty regarding the expert opinions using FANP	Wang et al. [81]
	Human resource in urban railway	Illustrating the superiority of FANP over FAHP in overlooking dependency and its ability to estimate it without considering hierarchical levels	Rezaie et al. [82]
Cargo loading operation		Applying DEMATEL to determine interdependency in experts' consensus and weighting EPCs based on their degree of similarity	Wang et al. [83]
Overhead crane operation		Proposing an ANP-DEMATEL model for causality map of parameters affecting PSFs and quantifying inter-dependency among them	Nurdiawati et al. [84]
Vital variables related to the situation awareness		Modeling inter-interactions among important variables in situation awareness fuzzy DEMATEL as well as reducing subjectivity uncertainty	Mohammadfam et al. [85]
Optimal reliability in hospital		Presenting FANP application to pairs compare of variables and handling dependency between EPCs in the expert consensus process	Abbaspour et al. [5]

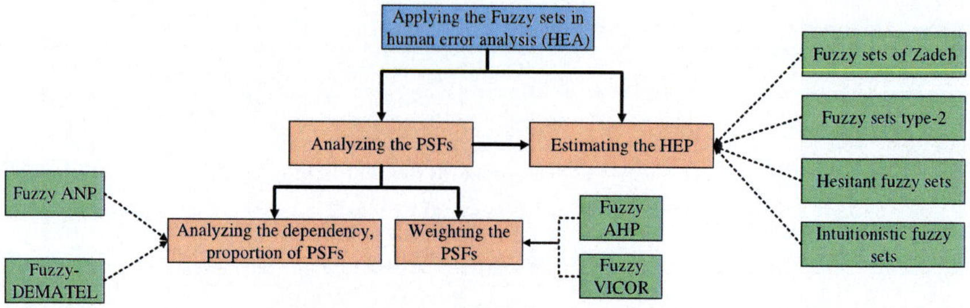

Fig. 5.4 An overview of FST and fuzzy MCDM methods in human error analysis

have a single vector (for example, the Simple Multi-attribute Ranking Technique or a complete matrix (for example, AHP). Although using a vector for input data is very data-driven and time-efficient, the main disadvantage of single-vector methods is that they cannot test the consistency of the pairwise comparisons presented. On the other hand, although the use of a complete matrix allows the stability of pairwise comparisons to be examined, if the decision-maker is faced with many questions, there is confusion in the calculations and as a result, the probability of making errors in the calculations has been increased. But in BWM, the decision-maker can perform unique calculations due to having multiple optimal paths and using two vectors [75]. In addition, the findings of the study by Nemeth et al. [63] show that the increasing complexity of the criteria weighting method can be associated with a lower probability for biased findings and a higher probability for practical challenges. Thus, the selection of the most appropriate technique for a decision problem should be based on accurate tolerance, given the shift between the complexity of the questions, the size of the decision group, and the resources available to build the multi-criteria decision analysis (MCDA) framework. The choice between these methods can be done based on several specific considerations such as simplicity, number of criteria, or strength of theoretical foundations of methods. In general, pre-testing questionnaires, reducing the number of questions that decision-makers have to answer, and articulating research findings are successful strategies in this area. Nemeth et al. [63] also argued that the Simple Multi-attribute Ranking Technique, combined with the rotational weighting technique and the AHP method, could reconcile complexity with the possibility of bias, and is probably the most feasible approach.

5.3.4 *HEP in Quantitative Risk Analysis (QRA)*

As mentioned earlier, HRA is one of the most important elements of probabilistic risk analysis in particularly in system safety or risk analysis. There are increasing requirements for regulators in all industries to conduct formal safety and health assessments. As part of these assessments, risk and reliability experts are required to assess human reliability in addition to performing hardware systems analysis

[89]. An examination of major historical events (e.g. Piper Alpha, Feyzin, Bhopal, and Texas City) has shown that human error has been implicated as both direct or indirect cause [90]. In this case, experts use quantitative risk assessment approaches and tools to incorporate human and hardware errors simultaneously. For example, they use cause-consequences analysis tools such as the Fault Tree Analysis (FTA), Bowtie, Bayesian Network, etc. [91–95].

Table 5.19 Application of human error analysis in QRA using fuzzy set theory

Application	Study description	Contribution	Reference
Vehicle leaved road	Quantifying the probabilities related to human error in FTA	Applicability of fuzzy sets in the absence of specialized information in accident investigation	Wang et al. [91]
Traffic operation	Investigating the effects of human error on accident risk in in-depth analysis	Applicability of fuzzy sets in calculating risk with low uncertainty in complex systems based on past incident information	Wang et al. [98]
System generation tube rapture in the nuclear industry	Investigating the importance of human error risk in the system based on FHERAM index	The ability of fuzzy sets to overcome ambiguity and uncertainty in the concept of human error in the system by considering recovery factors	Peng-cheng et al. [37]
γ-ray irradiator in the nuclear industry	Quantifying the probabilities related to human error in FTA	The ability of IFSs to quantify ambiguous human parameters and overcome subjectivity	Castigalia et al. [99]
General	Investigating the effect of human error in risk factor analysis	Applicability of fuzzy reasons model in collecting and quantifying expert opinions	Bouharati et al. [100]
High-speed railway	Quantification of probabilities related to human error in BN	The ability of Intuitionistic fuzzy to quantify ambiguous human parameters and overcome subjectivity	Ren et al. [101]
Process system	Assessing the system safety risk by considering human factors	Using fuzzy sets to overcome the ambiguity of crisp probability concepts in the BN	Zarei et al. [92]

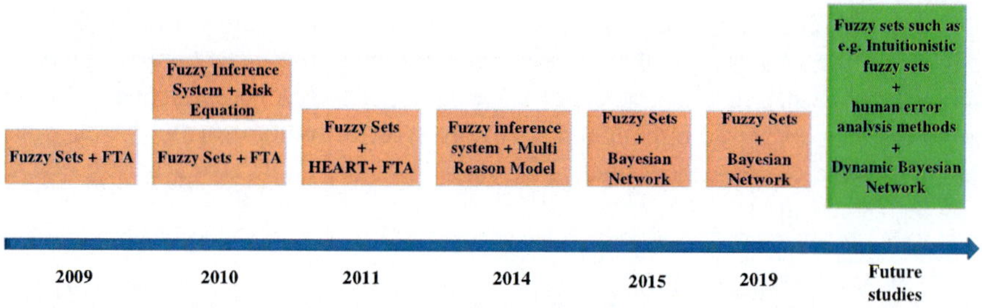

Fig. 5.5 The most popular hybrid models to analyze human error from 2009 to 2021

Table 5.19 presents details of the studies which tried to incorporate human error into QRA using fuzzy set theory in different domains. Each of these studies has its own benefits, but specialized studies show the superiority of the fuzzy-BN combination over other methods due to its lower uncertainty [92]. In this combination, HRA can be performed in a fuzzy framework, and then the fuzzy outputs can be entered into the BN. Then, the desired risk can be quantitatively calculated in a time-varying framework [96, 97].

Figure 5.5 shows the progress of the approaches and methods used concerning the inclusion of human error in the QRA Studies in this area began with the use of the fuzzy FTA in QRA so that first the accident model in the study of Wang et al. [91] with the top-to-down approach in the FTA was determined and then by combining fuzzy sets and analytical statistics (without using the experts' opinion), data quantification was performed to obtain a single quantitative risk. In another study, Wang et al. [98] stated that the main problem of FTA is its inability to dynamize the accident model, and their solution to overcome this problem was to integrate fuzzy sets and analytical statistics to create a single accident model. Thus, by combining information theory ideas with fuzzy sets and probability statistics, a new accident tree method was proposed that include the FTA and the mathematical simulation approach for QRA using a case study (off-road vehicle). In these two studies, they did not directly examine human error analysis but quantified human defects along with hardware defects in the FTA. Quantitative risk of human error has also been specifically calculated in studies. Peng-cheng et al. [37] calculated the numerical risk of human error specifically without the use of specialized tools. They used the three factors of human error probability, human error impact factor, and severity of the impact of human error for this purpose and, finally calculated the quantitative risk using a fuzzy inference system. Although FTA is known as a good way to assess quantitative risks, one of its main problems is the lack of categorization of basic events related to human error. It is therefore very useful if this method is combined with specific methods of human error analysis. This approach was developed in 2011 by Castigalia et al. [99]. They first categorized the technical defects and then identified the possible errors in the activities related to the γ-ray irradiator using the HEART method. Then they used fuzzy sets type-2 to quantify the errors. Human error can be influenced by factors with positive and negative effects. The multi-reason (e.g. multi-layer) model was another model in this regard that was used by Bouharati et al. [100] in 2014. In this study, this model was used in both quantitative risk assessment which was calculated by integrating with fuzzy sets, and by determining the factors related to each layer and entering different modes into the fuzzy inference system. Another shortcoming of FTA is the lack of nonlinear relationships between the parameters affecting human and technical errors. Therefore, using a flexible tool, such as a Bayesian network, can improve the accuracy of calculations. This approached was proposed in 2015 by Ren et al. [101]; where a Bayesian structure related to high-speed railway accidents was formed and its conditional probabilities were calculated using the fuzzy sets type-2.

A more complete sample of the fuzzy-Bayesian network (FBN) can be seen in the study of Zarei et al. [92] where human error was considered as one of the main causes (intermediate event) of natural gas leakage in the process industry and calculated the probabilities of basic events via two methods of crisp and fuzzy sets. The findings of their study showed that the FBN approach provides more accurate, realistic results with lower uncertainty than crisp probabilities. The reason for this superiority can be attributed to more accurate estimates of the aggregated views of experts concerning the rate of human error defects. Incidents usually occur in the context of human-technical interactions. Therefore, using an approach that can consider these interactions in a precise framework and provide more realistic numerical outputs is recognized as the best possible method. The FBN approach can be one of the most appropriate approaches in future studies if it can enter human errors into dynamic risk calculations using specialized methods such as HEART, CREAM, HFACS, etc.

Bowtie diagram is one of the most widely used methods in QRA. Mapping this method into the Bayesian network can provide a proper picture of the participation of human errors in accidents by separating human failures from technical failures. Quantification of the basic events related to the Bow tie in the process safety assessment at a natural gas station using fuzzy sets is shown here as a good example of the application of these sets [92]. Table 5.20 shows the basic events related to the human failures extracted from BT.

After collecting the primary data, fuzzy computations are performed as follows:

Table 5.20 The basic events related to the human failures in natural gas stations [92]

Basic event ID	Description	Basic event ID	Description
BE1	Wrong hazard assessment	BE7	Fear of explosion and fire
BE2	Intentional error	BE8	Poor/lack of equipment
BE3	Inadequate training	BE9	Permit implementations lack
BE4	Inadequate experience	BE10	Permit implementation failure
BE5	Failure in station restart up	BE11	Wrong permit
BE6	Interruption of gas subscribers	BE12	Without permit

1. Expert elicitation

To quantify the probability of basic events, a questionnaire was first sent to 3 experts and they were asked to comment on the probability of their occurrence. Table 5.21 shows the fuzzy linguistic terms in this regard.

2. Determining the weight of experts

In this step, to reduce the uncertainty, the weight of each expert was estimated based on the classification of Table 5.22 and Eqs. 5.17 and 5.18. The criteria for this estimate were professional position, experience time (year), education level, and age (year).

$$\begin{aligned}\text{Weight score of Expert i} = &\ \text{Score of PP of expert i} + \text{Score of ET of expert i} \\ &+ \text{Score of EL of expert i} + \text{Score of A of expert i}\end{aligned} \tag{5.17}$$

Table 5.21 Linguistic terms and fuzzy numbers to describe the likelihood of an event [92]

Linguistic terms	Description	Fuzzy sets (Scale 7)
Very High (VH)	Once in a month	(0.8, 1, 1, 1)
High-Very High (H-VH)	Once in every 1 to 3 months	(0.7, 0.9, 1, 1)
High (H)	Once in every 3 to 6 months	(0.6, 0.8, 0.8, 1)
Fairly High (FH)	Once in every 6 to 12 months	(0.5, 0.65, 0.65, 0.8)
Medium (M)	Once in every 1 to 5 years	(0.3, 0.5, 0.5, 0.7)
Fairly Low (FL)	Once in every 5 to 10 years	(0.2, 0.35, 0.35, 0.5)
Low (L)	Once in every 10 to 15 years	(0, 0.2, 0.2, 0.4)
Low-Very Low (L-VL)	Once in every 15 to 20 years	(0, 0, 0.1, 0.3)
Very Low (VL)	During life cycle of the system has not occurred/observed	(0, 0, 0, 0.2)

Table 5.22 Weighting score of different experts [92]

Constitution	Classification	Score	Constitution	Classification	Score
Professional position (PP)	Senior academic	5	Education level (EL)	PhD Master	5
	Junior academic	4		Bachelor	4
	Engineer	3		Higher National)	3
	Technician	2		Diploma (HND	2
	Worker	1		School level	1
Experience time (year) (ET)	≥ 30	5	Age (year) (A)	≥ 50	4
	20–29	4		40–49	3
	10–19	3			
	6–9	2		30–39	2
	≤ 5	1		< 30	1

$$\text{Weight factor of expert i} = \frac{\text{Weight score of expert i}}{\left(\sum_{i=1}^{n}\text{Weight score of expert i}\right)} \quad (5.18)$$

where PP is a professional position, ET is experience time, EL is education level, and A is Age of the expert.

For example, let assume that experts had the following conditions:

Weight score of Expert#1: PP: Senior academic, ET: 15 years, EL: Ph.D., A: 36 years = 5 + 3 + 5 + 2 = 15.
Weight score of Expert#2: PP: Engineer, ET: 20 years, EL: Bachelor, A: 43 years = 3 + 4 + 3 + 3 = 13.
Weight score of Expert#3: PP: Junior academic, ET: 9 years, EL: Master, A: 33 years = 4 + 2 + 4 + 2 = 12.

Based on these, the weight factor of each expert is:

Weight score of Expert#1: 15/ (15 + 13 + 12) = 0.375.
Weight score of Expert#2:13/ (15 + 13 + 12) = 0.325.
Weight score of Expert#3:12/ (15 + 13 + 12) = 0.300.

3. Estimating the fuzzy possibility using Eq. 5.19

$$M_i = \sum_{i=1}^{m} W_j A_{ji},\ j = 1, 2, \ldots, n \quad (5.19)$$

where M_i is the "fuzzy failure possibility" presenting the aggregated fuzzy value of the event i, W_j is the weighting score of Expert j, and A_{ij} is the linguistic value obtained from the expert j about the event i. m is the total number of events while n is the total number of experts. For example, let us assume that experts were commented on the "BE2: Intentional error" as linguistic terms in Table 5.23.

In this case, fuzzy possibility of BE2 is:

$$M_i = 0.375 \times (0.2, 0.35, 0.35, 0.5) + 0.325 \times (0, 0.2, 0.2, 0.4) + 0.300 \times (0.2, 0.35, 0.35, 0.5) = (0.135, 0.301, 0.301, 0.467)$$

Table 5.23 Linguistic terms related to the BE2 are based on the expert's opinion

Failure	Expert 1	Expert 2	Expert 3
BE2: Intentional error	FL (0.2, 0.35, 0.35,0.5)	L (0,0.2,0.2, 0.4)	FL (0.2, 0.35, 0.35,0.5)

4. Defuzzification

Finally, the fuzzy possibility related to each event was calculated using Eq. 5.20 as below:

$$X^* = \frac{\int_{a_1}^{a_2} \frac{x-a_1}{a_2-a_1} x dx + \int_{a_2}^{a_3} x dx + \int_{a_3}^{a_4} \frac{a_4-x}{a_4-a_3} x dx}{\int_{a_1}^{a_2} \frac{x-a_1}{a_2-a_1} dx + \int_{a_2}^{a_3} dx + \int_{a_3}^{a_4} \frac{a_4-x}{a_4-a_3} dx}$$
$$= \frac{1}{3} \otimes \frac{(a_4 + a_3)^2 - a_4 a_3 - (a_1 + a_2)^2 + a_1 a_2}{(a_4 + a_3 - a_1 - a_2)} \quad (5.20)$$

In our case, defuzzification was conducted as:

$$X = \frac{1}{3} \times \frac{(a_4 + a_3)^2 - a_4 a_3 - (a_1 + a_2)^2 + a_1 a_2}{(a_4 + a_3 - a_1 - a_2)}$$
$$= \frac{(0.467 + 0.301)^2 - (0.467 \times 0.301) - (0.135 + 0.301)^2 + (0.135 \times 0.301)}{(0.467 + 0.301 - 0.135 - 0.301)}$$
$$= 1.469$$

5. Estimating the fuzzy probability using Eqs. 5.15 and 5.16.

In our case, the BE2:
$k = \left|((1 - 1.469)/1.469)^{1/3} \times 2.301\right| = 1.57$ and probability $= \frac{1}{10^{1.57}} = 0.0269$.
We would like to refer the enthusiastic readers to original study [92] for more details.

5.3.5 *Human Behavior (Factor) Model*

Human and organizational factors (HOF) are one of the important issues that require special attention within industries, and the main reason for this importance can be attributed to their key role in the integrity of the production process [22]. Human factors are the main cause of at least 40% of all abnormal events in industrial activities [102]. The study of the most catastrophic accidents in history also shows the impact of human factors in accidents well. For instance, in the Bhopal accident, one of the worst industrial disasters in history, a combination of operator error, poor maintenance, weakness in the initial alert system, poor understanding of existing risks, and poor safety management lead to the leakage of methyl isocyanate from reservoirs that killed more than 2500 people [103]. Moreover, human errors such as inexperience, inadequate training, defect in human communication, and the lack of implementation of safety procedures lead to the occurrence of the Piper Alpha explosion which killed 167 workers [104]. Hence, learning from large industrial events has led to the analysis of human behaviors in various industries to become one of the most important essential elements necessary to enhance the objectives of safety and reduce the rate of incidents.

Studies on human factors have been carried out in several industries. Liu et al. [105] validated human behaviors in military simulations using fuzzy-Petri net in a dynamic framework. The reason for using fuzzy-Petri nets in evaluating human behavior models is their graphic nature. Furthermore, fuzzy-Petri nets have complete formal mechanisms for modeling and examining the properties of synchronized structures, and this formalism can be used to achieve AND/ OR parallelism associated with rule bases. In addition, strong and clear arguments of fuzzy-Petri nets can make fuzzy rule base arguments easy and understandable, and as a result, when the rules become petri nets, the argument based on the rule can be changed to the fuzzy Petri nets argument [105]. The fuzzy inference system was also used to predict human behavior and movements in industrial environments by Batchuluun et al. [106]. In this study, human movements were classified in terms of direction and anatomical type of movements in different fuzzy levels and then the type of human movements and behaviors were predicted based on the type of tasks and their requirements. As mentioned in the previous sections, the main advantage of IFSs over fuzzy sets type-1 and type-2 is the use of Euclidean geometric functions to determine the distance between non-membership and membership related to linguistic variables, as well as considering the consistency between specialists to overcoming the randomness. These advantages have enabled it to provide more accurate findings in combination with weighting methods. For example, Zarei et al. [21] used IFSs combined with AHP to weigh human behaviors in hot tapping operations on a natural gas pipeline and used its output for a Bayesian network. FBN findings are more accurate than BN findings because they have a better aggregation of expert opinions. In addition, the advantages of the best–worst method in weighting are discussed. Combining fuzzy sets with the Best Worst Method has also been one of the approaches used in the study of human behavior. In their two studies, Rostamabadi et al. [22, 107] used this approach to weighing the parameters affecting human behavior in the process industry. These studies can be considered from multiple aspects. First of all, it should be noted the significant superiority of the Best Worst Method over AHP in terms of interdependency. Secondly, we must pay attention to the fuzzy combination with this method to reduce subjectivity in using the opinions of experts. Third, the fuzzy Bayesian combination must also be considered. Fourth, the use of the Human factors analysis and classification (HFACS) technique to extract human behaviors in activities related to an ethylene storage tank and liquid hexane storage tank is also one of the tangible advantages of these studies over other studies. This combination, as can be seen in Fig. 5.5, has been an ideal approach because it has reduced computational uncertainty as much as possible.

HFACS, as one of the most popular human factor analysis methods, is selected to illustrate the application of fuzzy theory in human behavior analysis. HFACS examines active and latent failures in the organization and is based on Swiss cheese theory. This approach investigates the accidents at four levels of organizational impact, unsafe monitoring, preconditions for unsafe actions, and unsafe actions. 19 causal categories and at least 69 subgroups make up the framework of this method. Table

5.24 shows the identified failures in human operations in natural gas stations based on the HFAC framework (not all failures are shown in the table due to the large number of them in each sub-level. For details, please see [21]).

Table 5.24 The identified failures based on the HFAC framework [21]

HFACS level	Sub level	Failure symbol and its description
Organizational level (A)	Resource Management (A.1)	A.11. Failure in the maintenance of underground electrical equipment by Electricity Company (e.g., damaged insulation of electrical cables, fatigue and long life of cables)
	Organizational Climate (A.2)	A.21. The poor safety culture of the contractor, supervisory consultant and senior supervisors
	Operational Process (A.3)	A.31. Failure in process and procedure of permit to work (PTW)
Unsafe Supervision (B)	Inadequate Supervision (B.1)	B.11. Supervision failure in safety training of all personnel (contractor team, supervision consultant and senior supervisor) for excavation tasks especially related to underground electrical facilities
	Planned Inappropriate Operations (B.2)	B.21.Inappropriate scheduling by the contractor for operating drilling task on the day and consequently, failure to complete the operation during the shifts of the day
	Failed to Correct Known Problems (B.3)	B.31. Site manger's failure to correct unsafe behaviors related to safety issues
	Supervisory Violations (B.4)	B.41. Poor supervision for the implementation of safety regulations and requirements usually occurred by contractors such as failing to report unsafe acts, failing to PTW, failing to obtain a permit for night operation, and failing to contact the electrical company

(continued)

Table 5.24 (continued)

HFACS level	Sub level	Failure symbol and its description
Preconditions for Unsafe Acts (C)	Condition of Operators (C.1)	Because there are no occupational medical examinations (e.g., physical and mental) of workers. Thus, it is impossible to find the failure in physiological and mental conditions of workers
	Personnel Factors; Personnel resource management (C.2)	C.21. Failure in cooperation between an employer representative and site supervision
	Personnel Factors Fitness for Duty (C.3)	C.31. Lack of skills and qualifications of workers, HSE responsible person had misunderstanding of potential hazards of underground electrical facilities during excavation operation (failing in workers' skill acquisition process)
	Environmental Factors; Physical Environment (C.4)	There is no evidence reported for this item
	Environmental Factors; Technological Environment (C.5)	C.51. Insulation loss of electric cable which is related to night phase
Unsafe Act (D	Decision Errors (D.1)	D.14. Decision of contractor representative with supervision consultant according to oral cooperation for working at night shift
	Skill-based Errors (D.2)	There was no evidence reported for this item
	Perceptual (Knowledge) Errors (D.3)	D.31. Contact with night electrical cable without insulation due to lack of knowledge and information of day shift worker (He did not know the electrical wire at night would have a circuit)
	Routine Violations (D.4)	D.41. Lack of coordination with electricity company before start of the daily shiftwork
	Exceptional Violations (D.5)	D.51. Did not send night shift operation schedule for supervision consultant by contractor

(continued)

Table 5.24 (continued)

HFACS level	Sub level	Failure symbol and its description
	Sabotage (D.6)	There is no evidence reported for this item

The ultimate goal of HFACS is to identify failures related to human behavior at mentioned four levels. These failures are often estimated by subject matter experts' judgments which may be subject to ambiguity and vagueness. An Intuitionistic Fuzzy (IF) set is chosen to deal with this issue. After collecting the primary data, fuzzy computations are performed as follows:

For example, we find a defect in C.31.

C.31: Lack of skills and qualifications of workers, HSE responsible person had a misunderstanding of potential hazards of underground electrical facilities during excavation operation (failing in workers' skill acquisition process).

Table 5.25 shows the opinion of each expert about the C.31 failure.

Step1. Estimating a degree of similarity between the experts using Eq. 5.9. For example, to estimate a degree of similarity between expert2 and expert6 (S (E2&E6)):

$$S_{E1,E2} = 1 - 1/4(0.5 + 0.4 + 0.5 + 0.3) = 0.575$$

Based on it, the degrees of similarity (S) of all experts are provided in Table 5.26.

Table 5.25 The opinion of each expert about the C.31 failure [21]

Linguistic terms	Fuzzy numbers	Experts' opinions about A.13	
Very Low (VL)	(0, 0, 0.1, 0.2)	Expert 1: (VL)	(0, 0, 0.1, 0.2)
Low (L)	(0.1, 0.25, 0.25, 0.4))	Expert 2: (M)	(0.3, 0.5, 0.5, 0.7)
Medium (M)	(0.3, 0.5, 0.5, 0.7)	Expert 3: (VL)	(0, 0, 0.1, 0.2)
High (H)	(0.6, 0.75, 0.75, 0.9)	Expert 4: (M)	(0.3, 0.5, 0.5, 0.7)
Very High (VH)	(0.8, 0.9, 1, 1)	Expert 5: (VL)	(0, 0, 0.1, 0.2)
		Expert 6: (VL)	(0, 0, 0.1, 0.2)

Table 5.26 Degrees of similarity (S) of all experts

S(E1&E2)	0.575	S(E2&E4)	1.000	S(E3&E6)	1.000
S(E1&E3)	0.150	S(E2&E5)	0.575	S(E4&E5)	0.575
S(E1&E4)	0.575	S(E2&E6)	0.575	S(E4&E6)	0.575
S(E1&E5)	0.150	S(E3&E4)	0.575	S(E5&E6)	1.000
S(E1&E6)	0.150	S(E3&E5)	1.000	S(E3&E6)	1.000
S(E2&E3)	0.575	S(E2&E5)	0.575		

Table 5.27 The averages of agreement of all experts

AA (E1)	AA (E2)	AA (E3)	AA (E4)	AA (E5)	AA (E6)
0.32	0.66	0.66	0.66	0.66	0.66

Step 2: Estimating the average of agreement (AA) using Eq. 5.10. For instance, to estimate the average of agreement of expert1:

$$1/(6-1)(0.575+0.150+0.575+0.150+0.150)=0.32$$

The averages of agreement of all experts are shown in Table 5.27.

Step 3: Estimating the relative agreement using Eq. 5.11. For example, to estimating the relative agreement (RA) of expert1:

$$0.32\big/(0.32+0.66+0.66+0.66+0.66+0.66)=0.088$$

The relative agreements of all experts are provided in Table 5.28.

Step 4: Estimating the consensus coefficient using Eq. 5.12. For example, to estimating the consensus coefficient (CC) of expert1:

$$0.5\times 0.095+0.5\times 0.0883=0.0920$$

The consensus coefficients of all experts are shown in Table 5.29.

Step 5: Aggregation of experts' opinion using Eq. 5.13. For our case (C.31) we have:

$$\begin{aligned}&0.092\times(0,0,0.1,0.2)+0.232\times(0.3,0.5,0.5,0.7)\\&\quad+0.135\times(0,0,0.1,0.2)+0.175\times(0.3,0.5,0.5,0.7)\\&\quad+0.170\times(0,0,0.1,0.2)+0.197\times(0,0,0.1,0.2)\\&=(0.122,0.203,0.263,0.403)\end{aligned}$$

Step 6: The defuzzification based on center of gravity (COG) using Eq. 5.14. A computation for C.31 is:

Table 5.28 The relative agreements of all experts

RA (E1)	RA (E2)	RA (E3)	RA (E4)	RA (E5)	RA (E6)
0.088	0.182	0.182	0.182	0.182	0.182

Table 5.29 The consensus coefficients of all experts

CC (E1)	CC (E2)	CC (E3)	CC (E4)	CC (E5)	CC (E6)
0.092	0.232	0.135	0.175	0.170	0.197

Defuzzification of CC.31

$$= \frac{1}{3}\frac{(0.403 + 0.203)^2 - (0.403 \times 0.263) - (0.122 + 0.263)^2 + (0.122 \times 0.263)}{(0.403 + 0.203 - 0.122 - 0.263)}$$

$$= 0.656$$

Step 7: Finally, to converting the impact rate to the probability we use Eqs. 5.15 and 5.16. In our case, the probability of C.31 is:

$k = ((1/0.656) - 1)^{1/3} \times 2.301 = 1.85555$ and probability $= 1/\left(10^{1.8555}\right) = 0.0139$.

More information about the case study and employed methodology are available in the original study [21]

5.3.6 *Uncertainty Treatment*

The use of expert systems can help to improve the transparency and repeatability of results in risk analysis, especially when facing limited data. HRA, like other aspects of probabilistic risk analysis, has a degree of uncertainty [108]. Probabilistic calculations in risk assessment do not reflect the absolute reality of the system. Many definitions have been proposed for uncertainty considering the nature of different contexts. Simply, uncertainty refers to the degree to which calculations and estimates differ from reality or the measure of goodness of an estimate [109]. Figure 5.6 shows the different types of uncertainty in safety measurements and their sources. "Model Uncertainty" and "Completeness Uncertainty" depend on the expert's experience & knowledge and his/her attitude, while "Parameter Uncertainties" depend on the type of calculations and parameters of the study [110].

In HRA usually, two general classes of parameter uncertainty are as follows:

5.3.6.1 Stochastic or Aleatory Uncertainty

This uncertainty is caused by randomness and statistical variability [111] due to the inherent uncertainty in the parameter or event of interest. In simpler terms, it is due to the nature of probability analysis itself [112] and the probability distribution is the most common approach to deal with it [113]. A measure of how much is known about something (e.g., human error, error mode, PSFs) and thus a measure of uncertainty is the use of probability. Subjective probability is used to express our confidence in our knowledge of the event. This type of uncertainty cannot be eliminated or reduced, but it can be captured in our estimation using probabilistic approaches such as probability distribution and Monte Carlo simulation or Marko Chain-Monte Carlo simulation or Interval approaches which are constructive in conditions where we are sure regarding the bounds of a quantity, but we know little or nothing else.

5.3.6.2 Epistemic Uncertainty

This uncertainty often arises owing to lack of or insufficient knowledge or inconsistency in knowledge about the events, system states, processes, and phenomena that factor into the estimation which is common in HRA. It is much more evident in HRA compared to the other aspects of probabilistic risk analysis [113]. In general, incomplete understanding of human behavior, the mental phenomenon of human cognition, complex interactions affecting human behavior, and its unpredictability are the main reasons for the existence of epistemic uncertainty in HRA [30]. Fuzzy set theory and (fuzzy) evidential theory (which characterize the uncertainty caused by partial ignorance or incomplete knowledge) are the most common approaches to address this type of uncertainty in risk and reliability engineering.

One of the salient features of handling "uncertainty" is computations related to human capabilities. Depending on the type and adequacy of the information available, fuzzy sets can address uncertainty in three way [114].

1. Modeling with simple frequency equations: This modeling is used when there is sufficient historical information.
2. Modeling with statistical theories and Bayesian probabilities: This type of modeling is used when historical information is available but its amount is not enough.

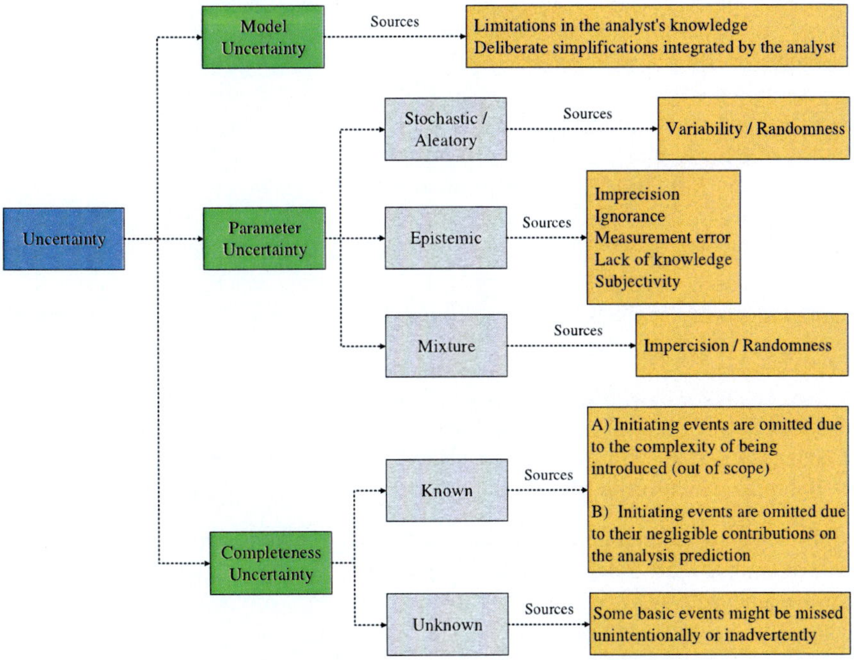

Fig. 5.6 The different types of uncertainty concerning human reliability analysis

3. Expert judgment: This solution is also employed when historical information is not available.

Uncertainty in HRA is due to incompleteness and confusion and can be modeled using membership functions. In this situation, uncertainty is influenced by inadequate statistical analytical methods, lack of sufficient information to analyze, the complexity of working conditions, and level of education and experience. It should be noted that the information of operators, as an essential source of information for analyzing tasks concerning system variables and their interrelationships, is ambiguous [114].

On the other hand, situational awareness as a variable and mental model is an important factor to improve performance and reduce human error, but it is difficult to model and evaluate [115]. This model can be qualitative (describing the interrelationships between a set of objects and events experienced) or quantitative (this model addresses questions such as "if it happens"), the second type of which can be represented by fuzzy sets [116].

In fuzzy sets, linguistic propositions are used as linguistic variables modeled with fuzzy numbers and fuzzy intervals [29]. Concerning HRA, the inputs and outputs of the "expert system" in the evaluation of PSFs are in the form of natural language (e.g. very unlikely, unlikely, probable, and very probable) [41]. This adaptation can reduce subjectivity, which is one of the most common causes of epistemic uncertainty. In fuzzy reliability, probability and binary modes are replaced by probability and fuzzy modes, and the error probability provides a more detailed description than the error probability in the probabilistic reliability [33].

Moreover, the main challenge is that evaluating membership functions and aggregating them always has a degree of uncertainty [9]. Few studies have specialized in assessing and evaluating uncertainty. In addition to the hypotheses considered in all studies, the fuzzy input uncertainty measure and the fuzzy input presence measure were used by Baraldi et al. [117] to measure and treat expert system uncertainty. Uncertainty assessment in these methods has an output that is displayed as FS-value. These values can be used to determine the accuracy of calculations. Comparative findings of treatment of uncertainty in BN and fuzzy expert systems used for a human reliability analysis by Baraldi et al. [118] also recommended the following two pathways for HRA studies:

- BN is preferred if the probability distribution is defined as describing the uncertainty of the specified parameters.
- Fuzzy expert system is preferred if there is a lack of knowledge and there are limitations in allocating probability distributions to describe uncertainty.

The event of human failure and dependence on human actions are other types of uncertainties that are treated with fuzzy logic [119]. Therefore, subjective uncertainty can be analyzed using FST [107, 120–122] or HEP correction with dependency considerations [60, 121] that have been developed for third-generation or dynamic HRA models concerning dependency handling [123].

5.4 Conclusion

In this chapter, the importance of fuzzy-driven human reliability investigations, as well as their various applications in different industries are reviewed. The findings revealed significant progress in human error analysis studies over the past decade. The results show that the fuzzy set theory can deal with subjectivity, handle the vague concepts in human error and factor modeling and thus address the epistemic uncertainty. Moreover, our findings presented the superiority of fuzzy ANP and fuzzy DEMATEL approaches over the fuzzy AHP. This superiority results from a strong capability for dependency modeling between PSFs in the two former techniques. Moreover, it seems that the fuzzy best-worth method has risen in popularity in recent years owing to its optimization and practical advantages. The authors found that the fuzzy-Bayesian network model when integrated into specific methods such as HEART, CREAM, HFACS can take significant benefits from both rigorous probabilistic reasoning and well-structured human factor or error modeling. In addition, BN-driven analysis is preferred in human reliability if the probability distribution can be properly defined to characterize the uncertainty. However, fuzzy experts' system is preferred if there is a lack of knowledge as well as limitations in allocating probability distributions to model uncertainty.

References

1. E. Zarei, F. Khan, R. Abbassi, Importance of human reliability in process operation: a critical analysis. Reliab. Eng. Syst. Saf. **211**, 107607 (2021)
2. E. Zarei, I. Mohammadfam, M.M. Aliabadi, A. Jamshidi, F. Ghasemi, Efficiency prediction of control room operators based on human reliability analysis and dynamic decision-making style in the process industry. Process Saf. Prog. **35**(2), 192–199 (2016)
3. M. Havlikova, M. Jirgl, Z. Bradac, Human reliability in man-machine systems. Procedia Engineering. **100**, 1207–1214 (2015)
4. R. Islam, R. Abbassi, V. Garaniya, F. Khan, Development of a human reliability assessment technique for the maintenance procedures of marine and offshore operations. J. Loss Prev. Process Ind. **50**, 416–428 (2017)
5. A. Abbaspour, M. Saremi, A. Alibabaei, P.S. Moghanlu, Determining the optimal human reliability analysis (HRA) method in healthcare systems using Fuzzy ANP and Fuzzy TOPSIS. Journal of Patient Safety and Risk Management. **25**(3), 123–133 (2020)
6. M. Philippart, Human reliability analysis methods and tools, in *Space Safety and Human Performance* (Elsevier, 2018), pp. 501–568
7. X. Zhou, X. Deng, Y. Deng, S. Mahadevan, Dependence assessment in human reliability analysis based on D numbers and AHP. Nucl. Eng. Des. **313**, 243–252 (2017)
8. A. Angelopoulou, K. Mykoniatis, N.R. Boyapati, Industry 4.0: the use of simulation for human reliability assessment. Procedia Manuf. **42**, 296–301 (2020)
9. B. Kim, R.R. Bishu, Uncertainty of human error and fuzzy approach to human reliability analysis. Int. J. Uncertain Fuzziness Knowledge-Based Syst. **14**(01), 111–129 (2006)
10. X. Deng, W. Jiang, Dependence assessment in human reliability analysis using an evidential network approach extended by belief rules and uncertainty measures. Ann. Nucl. Energy **117**, 183–193 (2018)

11. V.G. Krymsky, F.M. Akhmedzhanov, Assessment of human reliability under the conditions of uncertainty: SPAR-H methodology interpreted in terms of interval-valued probabilities. ASCE-ASME J. Risk Uncertainty Eng. Syst. Part B: Mech. Eng. **7**(2), 020907 (2021)
12. K.M. Groth, A. Mosleh, Deriving causal Bayesian networks from human reliability analysis data: a methodology and example model. Proc. Inst. Mech. Eng. Part O J. Risk Reliab. **226**(4), 361–379 (2012)
13. L. Mkrtchyan, L. Podofillini, V.N. Dang, Bayesian belief networks for human reliability analysis: a review of applications and gaps. Reliab. Eng. Syst. Saf. **139**, 1–16 (2015)
14. M.R. Martins, M.C. Maturana, Application of Bayesian Belief networks to the human reliability analysis of an oil tanker operation focusing on collision accidents. Reliab. Eng. Syst. Saf. **110**, 89–109 (2013)
15. L. Podofillini, V.N. Dang, A Bayesian approach to treat expert-elicited probabilities in human reliability analysis model construction. Reliab. Eng. Syst. Saf. **117**, 52–64 (2013)
16. M. Konstandinidou, Z. Nivolianitou, C. Kiranoudis, N. Markatos, A fuzzy modeling application of CREAM methodology for human reliability analysis. Reliab. Eng. Syst. Saf. **91**(6), 706–716 (2006)
17. M. Marseguerra, E. Zio, M. Librizzi, Human reliability analysis by fuzzy "CREAM." Risk Anal. Int. J. **27**(1), 137–154 (2007)
18. Q. Zhou, Y.D. Wong, H.S. Loh, K.F. Yuen, A fuzzy and Bayesian network CREAM model for human reliability analysis—the case of tanker shipping. Saf. Sci. **105**, 149–157 (2018)
19. Q. Lin, D. Wang, Facility layout planning with SHELL and Fuzzy AHP Method Based on human reliability for operating theatre. J. Healthcare Eng. **2019** (2019)
20. N. Wang, X. Du, M. Zhang, C. Xu, X. Lu, An improved weighted fuzzy CREAM model for quantifying human reliability in subway construction: modeling, validation, and application. Human Factors Ergon. Manuf. Serv. Ind. **30**(4), 248–265 (2020)
21. E. Zarei, M. Yazdi, R. Abbassi, F. Khan, A hybrid model for human factor analysis in process accidents: FBN-HFACS. J. Loss Prev. Process Ind. **57**, 142–155 (2019)
22. A. Rostamabadi, M. Jahangiri, E. Zarei, M. Kamalinia, S. Banaee, M.R. Samaei, A novel fuzzy Bayesian network-HFACS (FBN-HFACS) model for analyzing human and organization factors (HOFs) in process accidents. Process Saf. Environ. Prot. **132**, 59–72 (2019)
23. A.M. Kumar, S. Rajakarunakaran, V.A. Prabhu, Application of Fuzzy HEART and expert elicitation for quantifying human error probabilities in LPG refuelling station. J. Loss Prev. Process Ind. **48**, 186–198 (2017)
24. L.A. Zadeh. Fuzzy sets. Fuzzy sets, fuzzy logic, and fuzzy systems: selected papers by Lotfi A Zadeh: World Scientific, pp. 394–432 (1996)
25. S.S. Rivera, P.A. Baziuk, J.E. NúñezMcLeod (eds.), Fuzzy uncertainties in human reliability analysis, in *Proceedings of the World Congress on Engineering* (2011)
26. A. Jones, A. Kaufmann, H.-J. Zimmermann, *Fuzzy Sets Theory and Applications* (Springer Science & Business Media, 2012)
27. L.W.Rook, Jr., *Reduction of Human Error in Industrial Production* (Sandia Labs, Albuquerque, N. Mex.(USA), 1962)
28. L.-X. Hou, R. Liu, H.-C. Liu, S. Jiang, Two decades on human reliability analysis: a bibliometric analysis and literature review. Ann. Nuclear Energy **151**, 107969 (2021)
29. P.A. Baziuk, S.S. Rivera, J. Núñez Mc Leod, Fuzzy human reliability analysis: applications and contributions review. Adv. Fuzzy Syst. **2016** (2016)
30. S. Abrishami, N. Khakzad, S.M. Hosseini: A data-based comparison of BN-HRA models in assessing human error probability: an offshore evacuation case study. Reliab. Eng. Syst. Saf. **202**, 107043 (2020)
31. R. Boring, Dynamic human reliability analysis: benefits and challenges of simulating human performance. Citeseer (2007)
32. T. Onisawa, A representation of human reliability using fuzzy concepts. Inf. Sci. **45**(2), 153–173 (1988)
33. T. Onisawa, Y. Nishiwaki, Fuzzy human reliability analysis on the Chernobyl accident. Fuzzy Sets Syst. **28**(2), 115–127 (1988)

34. L.A. Zadeh, The concept of a linguistic variable and its application to approximate reasoning—I. Inf. Sci. **8**(3), 199–249 (1975)
35. X. Chen, X. Liu, Y. Qin, An extended CREAM model based on analytic network process under the type-2 fuzzy environment for human reliability analysis in the high-speed train operation. Qual. Reliab. Eng. Int. **37**(1), 284–308 (2021)
36. M. Karthick, T.P. Robert, C.S. Kumar, HFACS-based FAHP implementation to identify critical factors influencing human error occurrence in nuclear plant control room. Soft. Comput. **24**(21), 16577–16591 (2020)
37. P.-c Li, G.-h Chen, L.-c Dai, Li Z, Fuzzy logic-based approach for identifying the risk importance of human error. Saf. Sci. **48**(7), 902–913 (2010)
38. E. Zio, P. Baraldi, M. Librizzi, L. Podofillini, V.N. Dang, A fuzzy set-based approach for modeling dependence among human errors. Fuzzy Sets Syst. **160**(13), 1947–1964 (2009)
39. A. Azadeh, A.H. Farmand, Z.J. Sharahi, Performance assessment and optimization of HSE management systems with human error and ambiguity by an integrated fuzzy multivariate approach in a large conventional power plant manufacturer. J. Loss Prev. Process Ind. **25**(3), 594–603 (2012)
40. S.K. Tyagi, M. Akram, Human reliability evaluation for offshore platform musters using intuitionistic fuzzy sets. IEEE Trans. Fuzzy Syst. **21**(6), 1115–1122 (2013)
41. M.M. Aliabadi, Human error analysis in furnace start-up operation using HEART under intuitionistic fuzzy environment. J. Loss Prevention Process Ind. **69**, 104372 (2021)
42. S.Y. Cho, Reliability analysis of systems using trapezoidal fuzzy neutrosophic sets. J. Knowl. Inf. Technol. Syst. **11**(3), 293–299 (2016)
43. S. Ayber, N. Erginel (eds.), Developing the neutrosophic Fuzzy FMEA method as evaluating risk assessment tool, in *International Conference on Intelligent and Fuzzy Systems* (Springer, 2019)
44. M. Kamal, U.M. Modibbo, A. AlArjani, I. Ali, Neutrosophic fuzzy goal programming approach in selective maintenance allocation of system reliability. Complex Intell. Syst. **7**(2), 1045–1059 (2021)
45. M. Yazdi, N.A. Golilarz, A. Nedjati, K.A. Adesina, *Intelligent Fuzzy Pythagorean Bayesian Decision Making of Maintenance Strategy Selection in Offshore Sectors*, eds. by C. Kahraman, S. Cebi, S. Cevik Onar, B. Oztaysi, A.C. Tolga, I.U. Sari. Intelligent and Fuzzy Techniques for Emerging Conditions and Digital Transformation. INFUS 2021. Lecture Notes in Networks and Systems, vol. 308. (Springer, Cham, 2022). https://doi.org/10.1007/978-3-030-85577-2_70
46. A. Kumar, M. Ram, Reliability analysis for environment systems using dual hesitant fuzzy set, in *Advanced Fuzzy Logic Approaches in Engineering Science* (IGI Global, 2019), pp. 162–73
47. M. Yazdi, Risk assessment based on novel intuitionistic fuzzy-hybrid-modified TOPSIS approach. Saf. Sci. **110**, 438–48 (2018). https://doi.org/10.1016/j.ssci.2018.03.005
48. N.E. Oz, S. Mete, F. Serin, M. Gul, Risk assessment for clearing and grading process of a natural gas pipeline project: an extended TOPSIS model with Pythagorean fuzzy sets for prioritizing hazards. Hum. Ecol. Risk Assess. Int. J. **25**(6), 1615–1632 (2019)
49. M. Yazdi, Acquiring and sharing tacit knowledge in failure diagnosis analysis using intuitionistic and pythagorean assessments. J. Fail. Anal. Prev. **19**(2), 369–386 (2019)
50. M. Yazdi, Footprint of knowledge acquisition improvement in failure diagnosis analysis. Qual. Reliab. Engng. Int. **35**, 405–422 (2019). https://doi.org/10.1002/qre.2408doi.org/10.1002/qre.2408
51. M. Yazdi, A. Nedjati, E. Zarei, R. Abbassi, A novel extension of DEMATEL approach for probabilistic safety analysis in process systems. Saf. Sci. **121**, 119–136 (2020)
52. F. Kutlu Gündoğdu, C. Kahraman, Spherical fuzzy sets and spherical fuzzy TOPSIS method. J. Intell. Fuzzy Syst. **36**(1), 337–352 (2019)
53. C. Kahraman, F.K. Gündoğdu, *Decision Making with Spherical Fuzzy Sets: Theory and Applications* (Springer Nature, 2020)
54. F.K. Gündoğdu, C. Kahraman, A novel fuzzy TOPSIS method using emerging interval-valued spherical fuzzy sets. Eng. Appl. Artif. Intell. **85**, 307–323 (2019)

55. S. Gupta, P. Kumar, N.C. Karmakar, S.K. Palei (eds.), Quantification of human error rate in underground coal mines—a fuzzy mapping and rough set based approach, in *2013 IEEE International Conference on Industrial Engineering and Engineering Management* (IEEE, 2013)
56. P. Kumar, S. Gupta, Y.R. Gunda, Estimation of human error rate in underground coal mines through retrospective analysis of mining accident reports and some error reduction strategies. Saf. Sci. **123**, 104555 (2020)
57. E. Akyuz, Quantitative human error assessment during abandon ship procedures in maritime transportation. Ocean Eng. **120**, 21–29 (2016)
58. S.-T. Ung, Evaluation of human error contribution to oil tanker collision using fault tree analysis and modified fuzzy Bayesian Network based CREAM. Ocean Eng. **179**, 159–172 (2019)
59. P. Erdem, E. Akyuz: An interval type-2 fuzzy SLIM approach to predict human error in maritime transportation. Ocean Eng. **232**, 109161 (2021)
60. D. Gertman, H. Blackman, J. Marble, J. Byers, C. Smith, The SPAR-H human reliability analysis method. US Nuclear Regulatory Commission **230**(4), 35 (2005)
61. Ghasemi F, Ghasemi A, Kalatpour O: Prediction of human error probability during the hydrocarbon road tanker loading operation using a hybrid technique of fuzzy sets, Bayesian network and CREAM. Int. J. Occup. Saf. Ergon., 1–11 (2021)
62. G. Odu, Weighting methods for multi-criteria decision making technique. J. Appl. Sci. Environ. Manag. **23**(8), 1449–1457 (2019)
63. B. Németh, A. Molnár, S. Bozóki, K. Wijaya, A. Inotai, J.D. Campbell et al., Comparison of weighting methods used in multicriteria decision analysis frameworks in healthcare with focus on low-and middle-income countries. J. Comparative Effect. Res. **8**(4), 195–204 (2019)
64. M.-C. Chiu, M.-C. Hsieh, Latent human error analysis and efficient improvement strategies by fuzzy TOPSIS in aviation maintenance tasks. Appl. Ergon. **54**, 136–147 (2016)
65. S.T. Ung, W.M. Shen, A novel human error probability assessment using fuzzy modeling. Risk Ana. Int. J. **31**(5), 745–757 (2011)
66. F. Castiglia, M. Giardina, E. Tomarchio, THERP and HEART integrated methodology for human error assessment. Radiat. Phys. Chem. **116**, 262–266 (2015)
67. J. Geng, S. Murè, G. Baldissone, G. Camuncoli, M. Demichela (eds.), Human error probability estimation in ATEX-HMI area classification: from THERP to FUZZY CREAM, in *Proceedings of International Conference on Chemical & Process Engineering (ICheaP12)*, Milan (Italy) (2015)
68. M. Abbassinia, O. Kalatpour, M. Motamedzade, A. Soltanian, I. Mohammadfam, Dynamic human error assessment in emergency using fuzzy Bayesian cream. J. Res. Health Sci. **20**(1), e00468 (2020)
69. M.P. Sinabariba, M. Ghifari, E. Muslim, B. Moch (eds.), Analysis of human error risk with human reliability methods in construction projects, in *IOP Conference Series: Materials Science and Engineering* (IOP Publishing, 2020)
70. J. Zhao, S. Zeng, J. Guo (ed.), Human error oriented stochastic hybrid automation for human system interaction, in *2016 Annual Reliability and Maintainability Symposium (RAMS)* (IEEE, 2016)
71. L. Swaanika, R. Sujatha, D. Nagarajan, Estimation of human error using fuzzy relation. Int. J. Innov. Technol. Explor. Eng. (IJITEE) **8**(8) (2019)
72. J.C. Williams, HEART—a proposed method for achieving high reliability in process operation by means of human factors engineering technology, in *Proceedings of a Symposium on the Achievement of Reliability in Operating Plant* (Safety and Reliability Society (SaRS), NEC, Birmingham, 1985)
73. Y. Kim, J. Park, W. Jung, I. Jang, P.H. Seong, A statistical approach to estimating effects of performance shaping factors on human error probabilities of soft controls. Reliab. Eng. Syst. Saf. **142**, 378–387 (2015)
74. S. Kang, P.H. Seong, Performance shaping factor taxonomy for human reliability analysis on mitigating nuclear power plant accidents caused by extreme external hazards. Ann. Nuclear Energy **145**, 107533 (2020)

75. J. Rezaei, A concentration ratio for nonlinear best worst method. Int. J. Inf. Technol. Decis. Mak. **19**(03), 891–907 (2020)
76. S. Kovacevic, L. Papic, G. Janackovic, S. Savic, The analysis of human error as causes in the maintenance of machines: a case study in mining companies. S. Afr. J. Ind. Eng. **27**(4), 193–202 (2016)
77. S.-T. Ung, V. Williams, H. Chen, S. Bonsall, J. Wang, Human error assessment and management in port operations using fuzzy AHP. Mar. Technol. Soc. J. **40**(1), 73–86 (2006)
78. J. Hu, L. Zhang, Q. Wang, B. Tian, A structured hazard identification method of human error for shale gas fracturing operation. Hum. Ecol. Risk Assess. Int. J. **25**(5), 1189–1206 (2019)
79. I.J.L. dos Santos, J.E. França, L.F.M. Santos, A.N. Haddad, Allocation of performance shaping factors in the risk assessment of an offshore installation. J. Loss Prevention Process Ind. **64**, 104085 (2020)
80. G.F. Can, E.K. Delice, An advanced human error assessment approach: HEART and AV-DEMATEL. Human Factors Ergon. Manuf. Serv. Ind. **30**(1), 29–49 (2020)
81. W. Wang, X. Liu, Y. Qin, A modified HEART method with FANP for human error assessment in high-speed railway dispatching tasks. Int. J. Ind. Ergon. **67**, 242–258 (2018)
82. F. Mahdi Rezaie, A.M. Fakoor Saghih, N. Motahari Farimani, A novel hybrid approach based on CREAM and fuzzy ANP to evaluate human resource reliability in the urban railway. J. Transp. Saf. Secur., 1–39 (2020)
83. W. Wang, X. Liu, S. Liu, A hybrid evaluation method for human error probability by using extended DEMATEL with Z-numbers: a case of cargo loading operation. Int. J. Ind. Ergon. **84**, 103158 (2021)
84. A.A. Nurdiawati, L. Handoko, H.N. Amrullah, D. Dermawan, M. Shah, F. Hamzah (eds.), Human error probability analysis of overhead crane operation in steel fabrication company using SLIM-DEMATEL-ANP method, in *MATEC Web of Conferences* (EDP Sciences, 2018)
85. I. Mohammadfam, M.M. Aliabadi, A.R. Soltanian, M. Tabibzadeh, M. Mahdinia, Investigating interactions among vital variables affecting situation awareness based on Fuzzy DEMATEL method. Int. J. Ind. Ergon. **74**, 102842 (2019)
86. M.O. Barrios, H.F. Jiménez, S.N. Isaza (eds.), Comparative analysis between ANP and ANP-DEMATEL for six sigma project selection process in a healthcare provider, in *International Workshop on Ambient Assisted Living* (Springer, 2014)
87. A. Azizi, B. Malekmohammadi, H.R. Jafari, H. Nasiri, V.A. Parsa, Land suitability assessment for wind power plant site selection using ANP-DEMATEL in a GIS environment: case study of Ardabil province, Iran. Environ. Monitor. Assess. **186**(10), 6695–6709 (2014)
88. G. Sakthivel, M. Ilangkumaran, A. Gaikwad, A hybrid multi-criteria decision modeling approach for the best biodiesel blend selection based on ANP-TOPSIS analysis. Ain Shams Eng. J. **6**(1), 239–256 (2015)
89. J.E. Skogdalen, J.E. Vinnem, Quantitative risk analysis offshore—human and organizational factors. Reliab. Eng. Syst. Saf. **96**(4), 468–479 (2011)
90. F.I. Khan, S. Abbasi, Major accidents in process industries and an analysis of causes and consequences. J. Loss Prev. Process Ind. **12**(5), 361–378 (1999)
91. W. Wang, X. Jiang, X. Ren, H. Tan, Q. Cao (eds.), A simplified construction model for system safety analysis and quantified risk assessment: a case study of vehicle leaved road, in *2009 4th International Conference on Computer Science & Education* (IEEE, 2009)
92. E. Zarei, N. Khakzad, V. Cozzani, G. Reniers, Safety analysis of process systems using Fuzzy Bayesian Network (FBN). J. Loss Prev. Process Ind. **57**, 7–16 (2019)
93. A. Nedjati, M. Yazdi, R. Abbassi, A sustainable perspective of optimal site selection of giant air-purifiers in large metropolitan areas. Environ. Dev. Sustain. (2021). https://doi.org/10.1007/s10668-021-01807-0
94. S. Adumene, M. Okwu, M. Yazdi, M. Afenyo, R. Islam, C.U. Orji, et al., Dynamic logistics disruption risk model for offshore supply vessel operations in Arctic waters. Marit. Transp. Res. **2**, 100039 (2021). https://doi.org/10.1016/j.martra.2021.100039
95. M. Yazdi, F. Khan, R. Abbassi, R. Rusli, Improved DEMATEL methodology for effective safety management decision-making. Saf. Sci. **127**, 104705 (2020). https://doi.org/10.1016/j.ssci.2020.104705

96. L. Zhang, X. Wu, Y. Qin, M.J. Skibniewski, W. Liu, Towards a fuzzy Bayesian network based approach for safety risk analysis of tunnel-induced pipeline damage. Risk Anal. **36**(2), 278–301 (2016)
97. L. Zhang, X. Wu, M.J. Skibniewski, J. Zhong, Y. Lu, Bayesian-network-based safety risk analysis in construction projects. Reliab. Eng. Syst. Saf. **131**, 29–39 (2014)
98. W. Wang, X. Jiang, S. Xia, Q. Cao, Incident tree model and incident tree analysis method for quantified risk assessment: an in-depth accident study in traffic operation. Saf. Sci. **48**(10), 1248–1262 (2010)
99. F. Castiglia, M. Giardina, Fuzzy risk analysis of a modern γ-ray industrial irradiator. Health Phys. **100**(6), 622–631 (2011)
100. S. Bouharati, P. Allag, M. Belmahdi, M. Bounechada, S. Boumaïza (eds.), Risk factors analysis using the fuzzyfication of reason's model, in *2014 1st International Conference on Information and Communication Technologies for Disaster Management (ICT-DM)* (IEEE, 2014)
101. D. Ren, W. Zheng, Quantitative analysis methodology of non-deterministic causal relationship in risk analysis. Int. J. Secur. Its Appl. **9**(8), 261–274 (2015)
102. M. Ahmad, M. Pontiggia, M.C. Leva (eds.), Integration of human and organizational factors with quantitative risk assessment based on accident investigation, in *Proceedings of the 49th ESReDA Seminar* (2015). https://wwwresearchgate.snet/publication/289985155
103. T.A. Kletz, Accident investigation: keep asking "why?" J. Hazard. Mater. **130**(1–2), 69–75 (2006)
104. H. Book, *Reducing Error and Influencing Behaviour, Health and Safety Guidance HSG48/HSG* (Health and Safety Executive (HSE), Great Britain, 2009)
105. F. Liu, M. Yang, P. Shi (eds.), Verification and validation of fuzzy rules-based human behavior models, in *2008 Asia Simulation Conference-7th International Conference on System Simulation and Scientific Computing* (IEEE, 2008)
106. G. Batchuluun, J.H. Kim, H.G. Hong, J.K. Kang, K.R. Park, Fuzzy system based human behavior recognition by combining behavior prediction and recognition. Expert Syst. Appl. **81**, 108–133 (2017)
107. A. Rostamabadi, M. Jahangiri, E. Zarei, M. Kamalinia, M. Alimohammadlou, A novel Fuzzy Bayesian Network approach for safety analysis of process systems; An application of HFACS and SHIPP methodology. J. Clean. Prod. **244**, 118761 (2020)
108. J. Petronijevic, A. Etienne, J.-Y. Dantan, Human factors under uncertainty: a manufacturing systems design using simulation-optimisation approach. Comput. Ind. Eng. **127**, 665–676 (2019)
109. J. Mullins, Y. Ling, S. Mahadevan, L. Sun, A. Strachan, Separation of aleatory and epistemic uncertainty in probabilistic model validation. Reliab. Eng. Syst. Saf. **147**, 49–59 (2016)
110. H. Abdo, *Dealing with Uncertainty in Risk Analysis: Combining Safety and Security* (Université Grenoble Alpes, 2017)
111. J. Hu, Q. Zhou, A. McKeand, T. Xie, S.-K. Choi, A model validation framework based on parameter calibration under aleatory and epistemic uncertainty. Struct. Multidiscip. Optim. **63**(2), 645–660 (2021)
112. A. Mishra, E. Ahmadisharaf, B.L. Benham, D.L. Gallagher, K.H. Reckhow, E.P. Smith, Two-phase Monte Carlo simulation for partitioning the effects of epistemic and aleatory uncertainty in TMDL modeling. J. Hydrol. Eng. **24**(1), 04018058 (2019)
113. Q. Wang, L. Zhang, J. Hu, An integrated method of human error likelihood assessment for shale-gas fracturing operations based on SPA and UAHP. Process Saf. Environ. Prot. **123**, 105–115 (2019)
114. H.-N. Cho, H.-H. Choi, Y.-B. Kim, A risk assessment methodology for incorporating uncertainties using fuzzy concepts. Reliab. Eng. Syst. Saf. **78**(2), 173–183 (2002)
115. S. Dekker, E. Hollnagel, Human factors and folk models. Cogn. Technol. Work **6**(2), 79–86 (2004)
116. T.B. Sheridan, *Telerobotics, Automation, and Human Supervisory Control* (MIT Press, 1992)
117. P. Baraldi, M. Librizzi, E. Zio, L. Podofillini, V.N. Dang, Two techniques of sensitivity and uncertainty analysis of fuzzy expert systems. Expert Syst. Appl. **36**(10), 12461–12471 (2009)

118. P. Baraldi, L. Podofillini, L. Mkrtchyan, E. Zio, V.N. Dang, Comparing the treatment of uncertainty in Bayesian networks and fuzzy expert systems used for a human reliability analysis application. Reliab. Eng. Syst. Saf. **138**, 176–193 (2015)
119. F. Vanderhaegen, S. Zieba, S. Enjalbert, P. Polet, A Benefit/Cost/Deficit (BCD) model for learning from human errors. Reliab. Eng. Syst. Saf. **96**(7), 757–766 (2011)
120. M. Yazdi, A. Nedjati, E. Zarei, R. Abbassi, A reliable risk analysis approach using an extension of best-worst method based on democratic-autocratic decision-making style. J. Clean. Prod. **256**, 120418 (2020)
121. E. Zarei, F.Khan, M. Yazdi, A dynamic risk model to analyze hydrogen infrastructure. Int. J. Hydrogen Energy **46**(5), 4626–4643 (2021)
122. M. Yazdi, N.A. Golilarz, K.A. Adesina, A. Nedjati, Probabilistic Risk Analysis of Process Systems Considering Epistemic and Aleatory Uncertainties. A Comparison Study. Int. J. Uncertainty, Fuzziness Knowledge-Based Syst. **29**, 181–207 (2021). https://doi.org/10.1142/S0218488521500098
123. R.L. Boring, D.D. Dudenhoeffer, B.P. Hallbert, B.F. Gore, Modeling human reliability analysis using MIDAS. Citeseer (2006)

Chapter 6
Emergency Decision Making Fuzzy-Expert Aided Disaster Management System

Check for updates

Kehinde Adewale Adesina, Mohammad Yazdi, and Mohsen Omidvar

6.1 Introduction

There is no doubt about the astronomical increase in the global vulnerability to natural disasters and some other related phenomena. Due to the catastrophically emergencies and disasters that have plagued the entire universe, increasing world population and scarce resources, the societies have been highly vulnerable to various hazards. Statistically, the number of people susceptible and affected by disasters have been on the increase, however, the want of evidence to support the rate of increase of the real events responsible for the disaster has left researchers with no option than a growing vulnerability to disasters. Disasters are seen as occurrences that results to damages, displacement, social and economic disruption, casualties, health impairment, deterioration of health services, facilities and other infrastructure to the extent that exceptional emergency interventions are required outside of the affected area [1]. It is obvious disasters are potential threats to existing and future sustainable development. Hence, it is imperative for all hands to be on the deck to curtail the menace of disasters. Emergency intervention response geared toward reducing the vulnerability is therefore nonnegotiable.

The tendency to reduce vulnerability depends on some numbers of coordinated actions. These include (i) policy enactment and execution (ii) vulnerability appraisal

K. A. Adesina
Industrial Engineering Department, Near East University, KKTC, Nicosia, North Cyprus, Turkey

M. Yazdi (✉)
Faculty of Engineering and Applied Science, Memorial University of Newfoundland, St. John's, NL A1B 3X5, Canada

M. Omidvar
Department of Occupational Health and Safety Engineering, Faculty of Health and Nutrition, Bushehr University of Medical Sciences, Bushehr, Iran

M. Yazdi, *Linguistic Methods Under Fuzzy Information in System Safety and Reliability Analysis*, Studies in Fuzziness and Soft Computing 414,
https://doi.org/10.1007/978-3-030-93352-4_6

to decipher potential problems and opportunities (iii) disaster prevention and mitigation to reduce the extent of susceptibility, and lastly emergency readiness to determine and provide timely and adequate response. This Chapter addresses the last but important aspect of disaster vulnerability reduction.

Disaster prevention and mitigation could help to put in place measures or policies that would help to prevent the hazards or lessen the aftermath of the disaster. Such measure include flood mitigation works, workable land-use act, enhanced building standards and codes, relocation from vulnerable areas and structures among many other interventions. However, emergency readiness involves emergency plans, contingencies, adequate training and education about likely disaster and mitigation measures, and constant appraisal of these aforementioned measures.

Recent notable damages include the Japanese earthquake and tsunami of 2011, Pakistani heat wave of 2015, North Korean floods of 2018. Six categories of natural disasters are geological, meteorological, biological, and environmental pollution, fire, and marine disasters [2]. Recently, high likelihood, serious severity, wide coverage and other rising cost of disasters have been employed to determine emergency solutions to disaster [3]. Other forms of disasters have been reported [4–7] of which timely intervention and long-term risk mitigation have been suggested as major keys [7]. In the cases peculiar to Iran as a country with frequent floods disasters, timely and pragmatic intervention has been identified as a trusted means of ameliorating the aftermath of the disaster [8–13].

However, the lack of information for decision-makers has brought in some level of uncertainties [10] thereby making it tasking to achieve reliable emergency decisions through traditional methods [11]. Therefore, multi-criteria decision-making (MCDM) tools have been proposed for selecting optimum opinion among a set of alternatives in the presence of conflicting criteria [12, 14–16]. The MCDM methods are operations research models providing solutions to multi-contradictory criteria for determining optimum alternative among different possible alternatives. Emergency decision-making defined as effective ways of managing emergency cases [17–21]. Many mathematical models for solving decision-making problems have employed Bayesian Network [8, 22], Game theory [8, 14], Markov decisions [15], and Fuzzy set theory [16–18]. This Chapter provides a hybrid decision-making tool for classified decision-making cases. Basically, emergency readiness should be a robust long-term action that can enhance the overall capacity and the capability of the Manager to efficiently and effectively cope with the management of disasters to restore order.

An integrated method comprising of Best worst method (BWM) [19], Z numbers [20], and Zero-sum game [21] are introduced to deal with an emergency decision-making problem, particularly natural disasters suck as floods. BWM possesses performance than AHP in terms of the consistent ratio, conformity, total deviation, and minimum violation; and its initially application was deployed for determining importance weight of the criteria [19, 23]. The second step involves taking experts opinions and confidence level as slated by Z numbers which is more powerful for quantifying linguistic terms in emergency cases, for revealing uncertainty and retaining the integrity of the original data [24–27].

At the last step, confrontational zero-sum game is adopted to estimate optimal solution for the emergency case. In this game permits zero collaboration between two players with gain or loss of one player is balanced by the gains or losses of the other player. Therefore, the total loss and gain of the players has been found to always tend to zero [28–30]. The main goal of these players is the maximization of profits within the emergency decision-making procedure.

The chapter contributes first a BWM-Game theory hybrid model to enhance reliable results and secondly provides ways of employing the hybrid technique for managing a real-time natural disaster. The proposed approach has not been utilized before for emergency disaster management; therefore, it is the first of its kinds.

6.2 Methodology

A new three-stages framework for making realistic decisions during emergency provided. These stages are computing the weight of the criteria, building groups of payoff matrix, and determining the best solution. The importance weight for the criteria are estimated using the BWM [19, 31], the payoff matrix is evaluated, and the reliability of expert opinion is determine using Z numbers. Optimum response for interventions and emergency disruptive events is recognized using the zero-sum game theory [11]. The following shortcomings of a single decision-making tools (i) different decision-making tools loss of leads times between alternatives' rankings, (ii) ranking of the alternatives significantly depend on the importance weights of each criterion [32, 33], (iii) uncertainties due to human qualitative judgements and incomplete preference relationships [34, 35], and (iv) there are more tools which can also be employed to obtain robust formulation of the problem [36, 37].

Let $A = \{A_1, A_2, A_3, \ldots, A_m\}$. be set of alternatives, which is known as a *strategy of decision-makers* for l number of the decision-makers, $DM = \{DM_1, DM_2, DM_3, \ldots, DM_l\}$ be set of criteria $C = \{C_1, C_2, C_3, \ldots, C_n\}$ as a *strategy of nature*, and $\omega_j = \{\omega_1, \omega_2, \omega_3, \ldots, \omega_n\}$. is the decision-makers' importance weight according to the quality of experts' profile. It falls into the condition $0 \le \omega_j \le 1$. and $\sum_{j=1}^{n} \omega_j = 1$. In this regard, each participating decision-maker expresses individual payoff opinions of A_i. concerning the criterion C_j underlying the concept of Z numbers. Besides, the l. payoff matrices $P_k = \left[p_{ij}^k\right]_{m \times n}$ $(k = 1, 2, \ldots, l)$ n derived, here $p_{ij}^k = [\mu_{A_{ij}^k}, \mu_{R_{ij}^k}]$, where $\mu_{A_{ij}^k}$ is the decision-makers' opinion over A_i, concerning the criterion C_j collected from decision-maker, $\mu_{R_{ij}^k}$ is the reliability of the decision-makers' opinion over A_i related to the criterion C_j collected from decision-maker l. Accordingly based on the results, a zero-sum game between the decision-makers and nature can be illustrated as $B = \{\text{decision makers' nature}, A, C, P\}$.

Step 1 Importance weight of criteria

The importance weights of criteria for BWM are computed according to [38]; supposing C_B and C_W describes the best and least importance criterion from n criteria. With recurse to a nine-scale structure stated literature, the best to other vector (BO) defined as $C_{BO}^k, k = 1, 2, 3, \ldots, l$ is computed as:

$$C_{BO}^k = \left(C_{B1}^k, C_{B2}^k, \ldots, C_{Bn}^k\right) \tag{6.1}$$

where C_{Bj}^k is the opinion of the C_B over the C_j, and $C_{BB} = 1$.

Hence, finding l which is the aggregate of best to others' vectors $C_{BO} = (C_{B1}, C_{B2}, \ldots, C_{Bn})$. $C_{BO} = (C_{B1}, C_{B2}, \ldots, C_{Bn})$ can be conducted as follows:

$$C_{Bj} = \frac{C_{Bj}^k}{l}, j = 1, 2, .., n \tag{6.2}$$

Similarly, determine the others to worst vectors $C_{OW}, k = 1, 2, 3, \ldots, l$ can be estimated by;

$$C_{OW}^k = \left(C_{1W}^k, C_{2W}^k, \ldots, C_{nW}^k\right) \tag{6.3}$$

where C_{jW}^k is the opinion of the C_j over the C_W., and $C_{WW} = 1$..

Hence, for aggregating others to worsts' vectors $C_{OW} = (C_{1W}, C_{2W}, \ldots, C_{nW})$, the following expression is employed

$$C_{jW} = \frac{C_{jW}^k}{l}, j = 1, 2, \ldots, n \tag{6.4}$$

For estimated the optimal weight criteria for BWM, the ratio of $\frac{W_B}{W_j}$ and $\frac{W_j}{W_W}$ is followed by $\frac{W_B}{W_j} = C_{Bj}$ and $\frac{W_j}{W_W} = C_{jW}$. Therefore, maximizing the value of $\left|\frac{W_B}{W_j} - C_{Bj}\right|$ and minimizing the value of $\left|C_{jW} - \frac{W_j}{W_W}\right|$ would be enough to satisfy the required conditions. Thence, the model can be optimized optimal weight of criteria as:

Model 1:

$$\text{Min max}\left\{\left|\frac{W_B}{W_j} - C_{Bj}\right|, \left|C_{jW} - \frac{W_j}{W_W}\right|\right\},$$

Subject to

$$\sum_{j=1}^{n} w_j = 1, j = 1, 2, \ldots n.$$

Re-establishment of Model 1 to Model 2:

$$\min \xi$$

Subject to

$$\left|\frac{W_B}{W_j} - C_{Bj}\right| \leq \xi,$$
$$\left|C_{jW} - \frac{W_j}{W_W}\right| \leq \xi,$$
$$\sum_{j=1}^{n} w_j = 1,$$
$$w_j \geq 0, j = 1, 2, \ldots n.$$

The optimal importance weights can be denoted as $w^* = \left(w_1^*, w_2^*, \ldots, w_n^*\right)$.

The consistency ratio also known consistency index is determined by based on the maximum value of ξ and are presented Table 6.1.

$$CR = \frac{\xi^*}{CI} \tag{6.5}$$

Better consistency should have smaller CR value. The acceptance limit for CR for the studies is put as $CR \leq 0.2$.

Step 2 Payoff matrix and Z numbers

Z numbers are determine according to Zadeh [39] using two strategies. A linguistic term is firstly opinionated according to a subject, and secondly term the confident of the experts to express specific opinions are illustrated. Four facets are explored to inherently leverage the effects of these two strategies. These are technique for experts' opinion of elicitation, conversion of the confidence level, weighted confidence level

Table 6.1 Consistency index

C_{BW}	CI
1	0.00
2	0.44
3	1.00
4	1.63
5	2.30
6	3.00
7	3.73
8	4.47
9	5.23

of the opinions volunteered, and transformation of asymmetrical fuzzy numbers to symmetrical.

Step 3 Best Solution-Zero-sum game

Supposing DM.1 and DM.2 are the two decision-makers with DM.1 consistently aims at maximizing his or her returns while DM.2 aims at minimizing his or her returns, The total payoff of DM.1 and DM.2 is equal to zero. Note that,

Definition 1 A zero-sum game involving two DMs (players) is defined as a five-tuple in the following:

$$A = \{DM.1, DM.2, \mathcal{C}_D, \mathcal{C}_N, P\} \tag{6.9}$$

where $\mathcal{C}_D = \left\{ c_i | c_i \in \mathcal{C}_1 \geq 0, \sum_{i=1}^{m} c_i = 1, i = 1, 2, 3, \ldots, m \right\}$ is the mixed-based strategy set of DM.1, $\mathcal{C}_N = \left\{ \tilde{c}_j | \tilde{c}_j \in \mathcal{C}_1 \geq 0, \sum_{j=1}^{n} \tilde{c}_j = 1, i = 1, 2, 3, \ldots, n \right\}$ is the mixed-based strategy set of DM.2 (player 2), and P denotes the payoff matrix of $DM.1$ (player 1).

Definition 2 The payoff matrix is defined according to Ma et al. [38]; For $(c_i, \tilde{c}_j) \in \mathcal{C}_1 \times \mathcal{C}_2$, for$i = 1, 2, 3, \ldots, m$, a strategy pair where c_i is the strategy that DM.1 chooses and $\tilde{c}_j$ is the strategy that DM.2 chooses. For$(c_i, \tilde{c}_j)$, let $P = (p_{ij})_{m \times n}$ denotes the payoff matrix based on$DM.1$. Therefore, the payoff matrix of DM.2 is equal to$-P$. Zero-sum games are frequently solved using the Minimax theorem which is based on the linear programming duality or Nash equilibrium [40]. To solve the zero-sum game, Neumann (1959) developed the min–max theorem. For a zero-sum game, where ν shows the expected value of DM.1, a strategy pair $(c_i^*, \tilde{c}_j^*)$ is the Nash equilibrium point of$A = \{DM.1, DM.2, \mathcal{C}_D, \mathcal{C}_N, P\}$.

Model 3:

$$min \frac{1}{\nu} = \sum_{i=1}^{m} c_i$$

Subject to.

$$\sum_{i=1}^{m} p_{ij} c_i \geq 1, \quad j = 1, 2, 3, \ldots, n$$

$$c_i \geq 0, \quad i = 1, 2, 3, \ldots, m$$

then Nash equilibrium strategies of DM.1 and DM.2 derivable are $c_i^* = (\nu c_1, \ldots, \nu c_i, \ldots, \nu c_m)$ and $\tilde{c}_j^* = (\nu \tilde{c}_1, \ldots, \nu \tilde{c}_i, \ldots, \nu \tilde{c}_n)$ respectively.

The expected values of strategies $c_i (i = 1, 2, 3, \ldots, m)$ are determined as;

$$\mathcal{G}_i = c_i^* \cdot \sum_{j=1}^{n} \left(p_{ij}\tilde{c}_j^*\right) \cdot w_j^* \quad \forall i = 1, 2, 3, \ldots, m \tag{6.10}$$

For $\mathcal{G}_i \neq 0$, the best solution to strategy c_i can be estimated using the maximum $\mathcal{G}_i$ value. Similarly, when $\mathcal{G}_i = 0$, the row vectors related to $\mathcal{G}_i \neq 0$ are separated from the matrix, and the estimation is reconducted using model 1 and model 2.

6.3 Application of Study

A March 21 2019 natural disaster of Gorgan in the Golestan province of Iran is adopted the case study to illustrate the model. For complete description and geographical location of the city, and the extent of damage including the causalities, please refer to [4, 12, 41]. This study employs different alternative routes as possible solution to the emergency at Gorgan.

The Nash equilibrium strategy shows $A_i = (0.390, 0.000, 0.000, 0.000)$ and $C_j = (0.000, 0.000, 0.000, 0.676, 0.000)$; v is obtained as 2.56 and 1.479. Therefore, the vector weights are $A_i^* = (1.000, 0.000, 0.000, 0.000)$ and $C_j^* = (0.000, 0.000, 0.000, 1.000, 0.000)$. The expected values of alternatives gotten from Eq. (6.10) are $\mathcal{G}_1 = 0.3$, $\mathcal{G}_2 = 0$, $\mathcal{G}_3 = 0$, and $\mathcal{G}_4 = 0$. Therefore, alternative A_1- aircraft emergency delivery of relief supplies is the optimum solution. Subsequently, by eliminating $\mathcal{G}_i \neq 0$, the procedure progresses giving the ranking of the alternatives as $A_1 \succ A_3 \succ A_4 \succ A_2$. This ranking shows that aircraft emergency relief supplies delivery alternative should be prioritized. Sensitivity analysis (SA) conducted according to Saltelli (2002) reveals that of the confidence level is critical for the decision-making elicitation procedure. These findings are in tandem similar reports [20, 25, 26, 42, 43].

A comparison analysis with TOPSIS Simple Average Method [27, 44–50] and the proposed method are presented in Figs. 6.1 and 6.2. It seems based on appropriate restrictions such as time, complexity, etc. decision-makers can rely on other methods as well. However, the reflection obtained means from the three selected models nullifies validity of such deduction in the selection of the optimal solution. Spearman correlation coefficient confirms the conformity with priority of methods as reflected. The Spearman rank correlation coefficient between "proposed methodology" and "TOPSIS" is 0.2, between "Proposed methodology" and "Simple Average Method" is 0.8, and finally between "TOPSIS" and "Simple Average Method" is 0.4. The weight of each method is obtained to be 1, 0.6, 1.2 respectively for the proposed methodology, TOPSIS, and simple average method. More so simple average method has the highest correlation, followed by proposed methodology and then TOPSIS with the lowest correlation. Meanwhile, the simplicity of computation that compares existing approaches indicate that the proposed technique of this study is complex and time consuming.

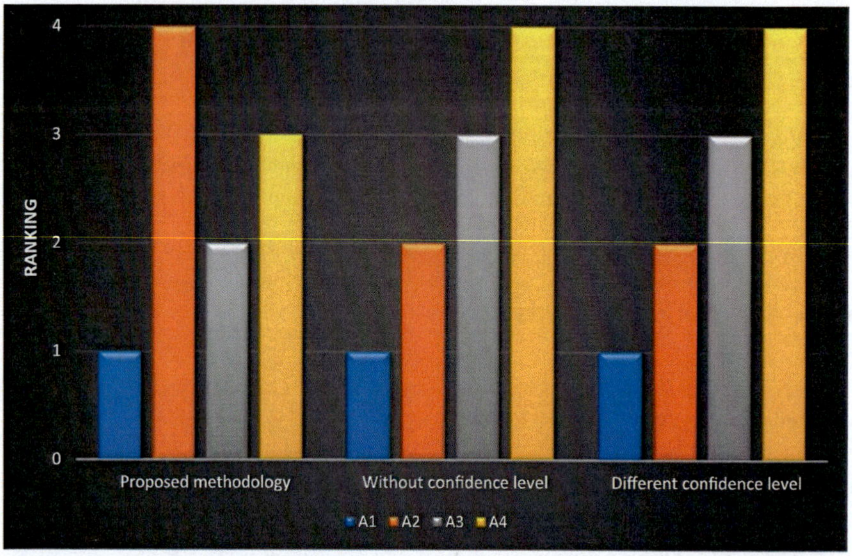

Fig. 6.1 Alternative ranking due to the sensitivity analysis

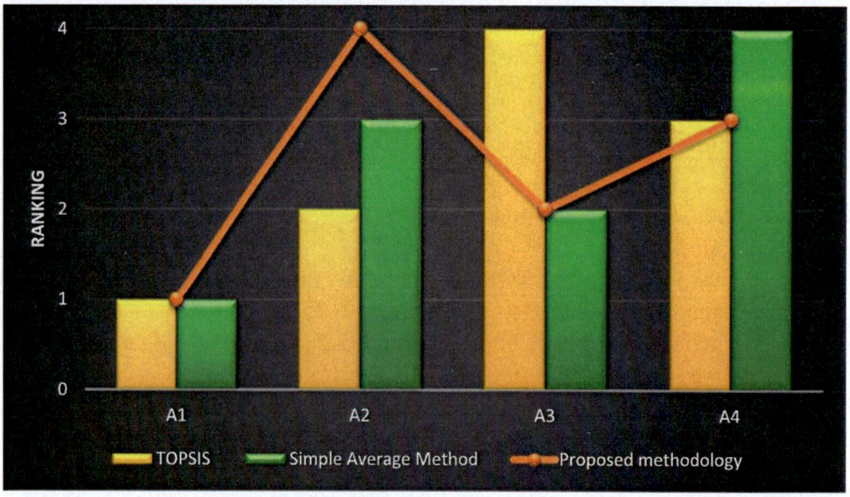

Fig. 6.2 The ranking based on different approaches

6.4 Implications

A novel horizon is being laid for the Country Crisis Management Organization (CCMO) of Iran by this study. The natural disaster is stochastic requiring a significant route. Therefore, it is pertinent to enhance the policies and emergency responses. To entrench a formidable procedure, group of decision-makers of diverse backgrounds should be engaged and be saddled with the responsibility identifying potential alternatives, criteria and to evaluate the applicability and effectiveness of the system. Furthermore, sources and the level of the risks of hazards should be ascertained especially those caused by humans. The emphasis will be placed on the (i) reduction in the probability of occurrence if hazards are human-based and (ii) reduction of the consequences of the event. Emergency response which includes order of the rapid response team, recourses, accessible passage and estimated havoc recked for all possible hazards needs to be conducted. The result of this study is of essence in that it confirms that performing risk-based decision-making models is essential in emergency decision-making process.

6.5 Conclusion

The Chapter presents novel method of combining BWM, Z numbers, and zero-sum game to deal with emergency decision-making problems in uncertain environments. The BWM obtained the importance weights of criteria, payoff matrix built by Z numbers and zero-sum game theory obtained the optimal solution. The flooding in Golestan, Iran is selected as a case study to demonstrate the feasibility and practicality of the proposed integrated approach. The results compared favorably with existing emergency decision-making models, this gives credence the propensity of the proposed model for assessing the risk of the disaster in such similar cases.

The study is therefore challenged by the problems of ignoring the effect of time. For studies, a probabilistic technique like Bayesian updating mechanism could be adopted. Additionally, a combination of Z-numbers with other types of fuzzy numbers such as Pythagorean, intuitionistic, to gage the accuracy of the opinion could bring new dimension to the output of the proposed model. Moreover, the influence of cooperative and psychosocial behavior of decision-makers on the output needs to be examined. Finally, the proposed approach could compare well with other well-known MCDM quality models such as VIKOR, MABAC, MARCOS, etc.

References

1. Bankoff, G. Frerks, D. Hilhorst, *Mapping Vulnerability: Disasters, Development and People*, London (2004)
2. L. Zhou, X. Wu, Z. Xu, H. Fujita, Emergency decision making for natural disasters: an overview. Int. J. Disaster Risk Reduct. **27**, 567–576 (2018). https://doi.org/10.1016/j.ijdrr.2017.09.037
3. The rising cost of catastrophes, Econ. (2012). https://www.economist.com/leaders/2012/01/14/the-rising-cost-of-catastrophes
4. M. Ebrahim, B. Nastaran, C. Timothy, Non-compensatory decision model for incorporating the sustainable development criteria in flood risk management plans. SN Appl. Sci. **2**, 1–11 (2020). https://doi.org/10.1007/s42452-019-1695-6

5. N. Chitsaz, M.E. Banihabib, Comparison of different multi criteria decision-making models in prioritizing flood management alternatives. Water Resour. Manage., 2503–2525 (2015). https://doi.org/10.1007/s11269-015-0954-6
6. W. Douven, J. Buurman, Planning practice in support of economically and environmentally sustainable roads in floodplains: the case of the Mekong delta floodplains. J. Environ. Manage. **128**, 161–168 (2013). https://doi.org/10.1016/j.jenvman.2013.04.048
7. J. Ni, L. Sun, T. Li, Z. Huang, A.G.L. Borthwick, Assessment of flooding impacts in terms of sustainability in mainland China. J. Environ. Manage. **91**, 1930–1942 (2010). https://doi.org/10.1016/j.jenvman.2010.02.010
8. M. Yazdi, N.A. Golilarz, A. Nedjati, K.A. Adesina, Intelligent Fuzzy Pythagorean Bayesian Decision Making of Maintenance Strategy Selection in Offshore Sectors. In: C. Kahraman, S. Cebi, S. Cevik Onar, B. Oztaysi, A.C. Tolga, I.U. Sari (eds). Intelligent and Fuzzy Techniques for Emerging Conditions and Digital Transformation. INFUS 2021. Lecture Notes in Networks and Systems, vol. 308. (Springer, Cham, 2022). https://doi.org/10.1007/978-3-030-85577-2_70
9. M. Li, P. Cao, Computers & Industrial Engineering Extended TODIM method for multi-attribute risk decision making problems in emergency response. Comput. Ind. Eng. **135**, 1286–1293 (2019). https://doi.org/10.1016/j.cie.2018.06.027
10. M. Yazdi, S. Kabir, M. Walker, Uncertainty handling in fault tree based risk assessment: State of the art and future perspectives. Process Saf. Environ. Prot. **131**, 89–104 (2019). https://doi.org/10.1016/j.psep.2019.09.003
11. L. Zhang, Y. Wang, X. Zhao, A new emergency decision support methodology based on multi-source knowledge in 2-tuple linguistic model. Knowledge-Based Syst. **144**, 77–87 (2018). https://doi.org/10.1016/j.knosys.2017.12.026
12. S. Daneshvar, M. Yazdi, K.A. Adesina, Fuzzy smart failure modes and effects analysis to improve safety performance of system: case study of an aircraft landing system. Qual. Reliab. Eng. Int., 1–20 (2020). https://doi.org/10.1002/qre.2607
13. F. Nawaz, M.R. Asadabadi, N.K. Janjua, O.K. Hussain, E. Chang, M. Saberi, An MCDM method for cloud service selection using a Markov chain and the best-worst method. Knowledge-Based Syst. **159**, 120–131 (2018). https://doi.org/10.1016/j.knosys.2018.06.010
14. J. Ding, J. Cai, G. Guo, C. Chen, An emergency decision-making method for urban rainstorm water-logging : a China study. Sustainability, 1–21 (2018). https://doi.org/10.3390/su10103453
15. M. Chen, Z. Dong, W. Jia, X. Ni, H. Yao, Multi-objective joint optimal operation of reservoir system and analysis of objectives competition mechanism: a case study in the upper reach of the Yangtze River. Water Artic. **11** (2019). https://doi.org/10.3390/w11122542
16. M. Yazdi, A perceptual computing-based method to prioritize intervention actions in the probabilistic risk assessment techniques. Qual. Reliab. Eng. Int., 1–27 (2019). https://doi.org/10.1002/qre.2566
17. M. Yazdi, Introducing a heuristic approach to enhance the reliability of system safety assessment. Qual. Reliab. Eng. Int., 1–27 (2019). https://doi.org/10.1002/qre.2545
18. S.P. Simonovic, S. Ahmad, Computer-based model for flood evacuation emergency planning. Nat. Hazards, 25–51 (2005)
19. J. Rezaei, Best-worst multi-criteria decision-making method. Omega (United Kingdom) **53**, 49–57 (2015). https://doi.org/10.1016/j.omega.2014.11.009
20. L.A. Zadeh, A note on Z-numbers. Inf. Sci. (Ny) **181**, 2923–2932 (2011). https://doi.org/10.1016/j.ins.2011.02.022
21. Y. Chen, M. Larbani, Two-person zero-sum game approach for fuzzy multiple attribute decision making problems. Fuzzy Sets Syst. **157**, 34–51 (2006). https://doi.org/10.1016/j.fss.2005.06.004
22. M. Yazdi, A review paper to examine the validity of Bayesian network to build rational consensus in subjective probabilistic failure analysis. Int. J. Syst. Assur. Eng. Manag. **10**, 1–18 (2019). https://doi.org/10.1007/s13198-018-00757-7
23. M. Mohammadi, J. Rezaei, Bayesian best-worst method: a probabilistic group decision making model. Omega (United Kingdom), 1–8 (2019). https://doi.org/10.1016/j.omega.2019.06.001

24. S. Vahedberdi, A. Kornejady, M. Ownegh, *Application of the Coupled TOPSIS—Mahalanobis Distance for Multi-hazard-Based Management of the Target Districts* (Springer, Netherlands, 2019). https://doi.org/10.1007/s11069-019-03617-0
25. R.A. Aliev, W. Pedrycz, O.H. Huseynov, Functions defined on a set of Z-numbers. Inf. Sci. **423**, 353–375 (2018). https://doi.org/10.1016/j.ins.2017.09.056
26. H. Aboutorab, M. Saberi, M.R. Asadabadi, O. Hussain, E. Chang, ZBWM: The Z-number extension of Best Worst Method and its application for supplier development. Expert Syst. Appl. **107**, 115–125 (2018). https://doi.org/10.1016/j.eswa.2018.04.015
27. S. Yousefzadeh, K. Yaghmaeian, A.H. Mahvi, S. Nasseri, N. Alavi, R. Nabizadeh, Comparative analysis of hydrometallurgical methods for the recovery of Cu from circuit boards: optimization using response surface and selection of the best technique by two-step fuzzy AHP-TOPSIS method. J. Clean. Prod. **249**, 119401 (2020). https://doi.org/10.1016/j.jclepro.2019.119401
28. Y. Koca, O. Muge, Solving two-player zero sum games with fuzzy payoffs when players have different risk attitudes. Qual. Reliab. Eng. Int., 1461–1474 (2018). https://doi.org/10.1002/qre.2322
29. J. Xu, J.Y. Dong, S.P. Wan, J. Gao, Multiple attribute decision making with triangular intuitionistic fuzzy numbers based on zero-sum game approach. Iran. J. Fuzzy Syst. **16**, 97–112 (2019)
30. J. Frigout, S. Tasseel-ponche, A. Delafontaine, Strategy and decision making in Karate. Front. Psychol. **10**, 1–9 (2020). https://doi.org/10.3389/fpsyg.2019.03025
31. A. Nedjati, M. Yazdi, R. Abbassi, A sustainable perspective of optimal site selection of giant air-purifiers in large metropolitan areas. Environ. Dev. Sustain (2021). https://doi.org/10.1007/s10668-021-01807-0
32. M. Yazdi, F. Khan, R. Abbassi, R. Rusli, Improved DEMATEL methodology for effective safety management decision-making. Saf. Sci. **127**, 104705 (2020). https://doi.org/10.1016/j.ssci.2020.104705
33. V. Ibáñez-Forés, M.D. Bovea, V. Pérez-Belis, A holistic review of applied methodologies for assessing and selecting the optimal technological alternative from a sustainability perspective. J. Clean. Prod. **70**, 259–281 (2014). https://doi.org/10.1016/j.jclepro.2014.01.082
34. K. Govindan, S. Rajendran, J. Sarkis, P. Murugesan, Multi criteria decision making approaches for green supplier evaluation and selection: a literature review. J. Clean. Prod. **98**, 66–83 (2015). https://doi.org/10.1016/j.jclepro.2013.06.046
35. M. Yazdi, N.A. Golilarz, K.A. Adesina, A. Nedjati, Probabilistic risk analysis of process systems considering epistemic and aleatory uncertainties: a comparison study. Int. J. Uncertainty, Fuzziness Knowl.-Based Syst. **29**, 181–207 (2021). https://doi.org/10.1142/S0218488521500098
36. W. Ingwersen, H. Cabezas, A. V Weisbrod, T. Eason, B. Demeke, X. Ma, T.R. Hawkins, S.-J. Lee, J.C. Bare, M. Ceja, Integrated metrics for improving the life cycle approach to assessing product system. Sustainability **6** (2014). https://doi.org/10.3390/su6031386
37. M. Yazdi, F. Khan, R. Abbassi, Microbiologically influenced corrosion (MIC) management using Bayesian inference. Ocean Eng. (2021). https://doi.org/10.1016/j.oceaneng.2021.108852
38. J. Ma, Y. Zheng, B. Wu, L. Wang, Equilibrium topology of multi-agent systems with two leaders. Automatica **73**, 200–206 (2016). https://doi.org/10.1016/j.automatica.2016.07.005
39. L.A. Zadeh, The concept of a linguistic variable and its application to approximate reasoning-I. Inf. Sci. (Ny). **249** (1975)
40. K. Binmore, *Playing for Real: A Text on Game Theory* (Oxford University Press, New York, 2007)
41. Iran International, Unprecedented Flood in North of Iran (2019)
42. R.A. Aliev, A.V. Alizadeh, O.H. Huseynov, The arithmetic of discrete Z-numbers. Inf. Sci. (Ny) **290**, 134–155 (2015). https://doi.org/10.1016/j.ins.2014.08.024
43. B. Kang, Y. Deng, K. Hewage, R. Sadiq, A method of measuring uncertainty for Z-number. IEEE Trans. FUZZY Syst. **27**, 731–738 (2019)
44. F.E. Boran, S. Genç, M. Kurt, D. Akay, A multi-criteria intuitionistic fuzzy group decision making for supplier selection with TOPSIS method. Expert Syst. Appl. **36**, 11363–11368 (2009). https://doi.org/10.1016/j.eswa.2009.03.039

45. H. Gupta, M.K. Barua, A framework to overcome barriers to green innovation in SMEs using BWM and Fuzzy TOPSIS. Sci. Total Environ. **633**, 122–139 (2018). https://doi.org/10.1016/j.scitotenv.2018.03.173
46. M. Behzadian, S. Khanmohammadi Otaghsara, M. Yazdani, J. Ignatius, A state-of the-art survey of TOPSIS applications. Expert Syst. Appl. **39**, 13051–13069 (2012). https://doi.org/10.1016/j.eswa.2012.05.056
47. S.C. Zhang, H. Wang, Z. Liu, S. Zeng, Y. Jin, T. Baležentis, A comprehensive evaluation of the community environment adaptability for elderly people based on the improved TOPSIS. Information **10** (2019). https://doi.org/10.3390/info10120389
48. D.A. Wood, Supplier selection for development of petroleum industry facilities, applying multi-criteria decision making techniques including fuzzy and intuitionistic fuzzy TOPSIS with flexible entropy weighting. J. Nat. Gas Sci. Eng. **28**, 594–612 (2016). https://doi.org/10.1016/j.jngse.2015.12.021
49. S. Sharma, S. Balan, An integrative supplier selection model using Taguchi loss function, TOPSIS and multi criteria goal programming. J. Intell. Manuf. **24**, 1123–1130 (2013). https://doi.org/10.1007/s10845-012-0640-y
50. S. Ramya, V. Devadas, Integration of GIS, AHP and TOPSIS in evaluating suitable locations for industrial development: a case of Tehri Garhwal district, Uttarakhand, India. J. Clean. Prod. **238**, 117872 (2019). https://doi.org/10.1016/j.jclepro.2019.117872

Chapter 7
Smart Decision Fuzzy-Based Data Envelopment Model for Failure Modes and Effects Analysis

Kehinde Adewale Adesina, Mohammad Yazdi, Esmaeil Zarei, and Mostafa Pouyakian

7.1 Introduction

Since the 60s, Failure Modes and Effects Analysis (FMEA) has been the means of assessing processes systematically and, proactively to study, detect, examine and control them for any apparent or inherent. This is done to redeem or salvage processes that are most in need of change. FMEA deciphers the reason for failure (failure modes), causes (failure causes), and the consequences of the failure (failure effects). Thus, by these FMEA assesses processes for possible failures and offers corrective actions to proactively forestall any adverse consequences of the release of the failure. This emphasizes preventive rather than reactive actions in remediating the risk to the facility, environment, and personnel. FMEA is applicable to both existing and new processes.

Furthermore, FMEA is widely used to define, identify, and eliminate potential failures and because of the risk assessment procedure is continues, it causes improvement in process designs [1–3]. According to evidence-based experts' opinions from the FMEA team members, each failure mode (FM) will be obtained risk propriety number (RPN) to represent the level of its risk. Subsequently, some recommendations are provided on how failure modes (FMs) might be controlled to prevent them

K. A. Adesina
Industrial Engineering Department, Near East University, KKTC, Nicosia, North Cyprus, Turkey

M. Yazdi (✉)
Faculty of Engineering and Applied Science, Memorial University of Newfoundland, St. John's, NL A1B 3X5, Canada

E. Zarei
Centre for Risk, Integrity and Safety Engineering (C-RISE), Faculty of Engineering and Applied Science, Memorial University of Newfoundland, St. John's, NL A1B 3X5, Canada

M. Pouyakian
Department of Occupational Health and Safety Engineering, School of Public Health and Safety, Shahid Beheshti University of Medical Sciences, Tehran, Iran

M. Yazdi, *Linguistic Methods Under Fuzzy Information in System Safety and Reliability Analysis*, Studies in Fuzziness and Soft Computing 414,
https://doi.org/10.1007/978-3-030-93352-4_7

from having an improper effect on the system. Three risk factors in FMEA including the severity of failure effect (S), occurrence of failure modes (O), and probability of not detection (D) are examined to compute the RPN scores for all potential FMs; in this regard, FMEA has enough capability to increase the reliability and safety performance of the system [4, 5]. So far, FMEA as one of the most powerful methods has been widely applied to many industries including manufacturing, aerospace, chemical, software engineering, marine and offshore, medical management, food industry, and nuclear [4–15]. However, the conventional FMEA technique has been widely criticized because of varieties of reasons in recent decades.

Some of the most important shortages of conventional FMEA which are reported in the literature are provided as follows [16]. (i) The weights of each risk factor are considered the same whereas according to the type of system it may not be highly necessary to be considered in this way. (ii) To determine all three risk factors for each FM is usually difficult for risk assessors. (iii) There is no evidence to show that the best mathematically way to compute the RPN score is the multiplication of three risk factors. (iv) Different combination of risk factors has high possibility to produce the same RPN score like as (10, 5, 6) and (3, 10, 10); however, their hidden might be different. (v) RPN only considered three risk factors S, O, and D as main parts of safety issues whereas other factors such as financial issues and reputation of the system may be important as well. (vii) RPN is computed in discrete ordinal scales of measure although production of three risk factors has not been meant on an ordinal scale. (viii) The production of three risk factors can only get 120 numbers out of 1000 meaning that RPN scores are not continuous and have many holes. (ix) The direct and indirect relationships between FMs and their causes are not taken to consideration. (x) The three risk factors are elicited from experts' knowledge; in this case, such information is commonly vague or uncertain.

Therefore, conventional FMEA cannot handle uncertainty and ambiguity during risk assessment. (xi) Experts prefer to express their opinions in qualitative terms; however, the relation between risk factor D and associated rating is linear, while the O and associated rating have a nonlinear relationship. (xii) The main mathematical form due to the production of RPN is intensely sensitive to the variation of risks. In other words, a small value of one of the risk factors is leading to many diverse effects on the RPN scores. (xiv) Conventional FMEA is only representing the measured risk using RPN and ignoring the importance of corrective actions; however, the effectiveness of corrective actions has a high necessity. (xv) In addition, sometimes FMs and even more provided corrective actions have dependency between each other, while conventional FMEA cannot recognize any dependency through the risk assessment procedure. To obtain comprehensive shortages of conventional FMEA and their frequency, one can refer to Appendix of [16].

However, in recent decades, the number of studies that attempted to cope with the shortages of conventional FMEA is highly considerable. For example, to deal with uncertainty, an extension of fuzzy set theory can be used like as [17–20], to deal with the dependency of corrective actions and FMs [8, 20], and using different weights of risk factors [21, 22]. Additionally, Data Envelopment Analysis (DEA) as an effective tool to evaluate the efficiency of the system which is introduced by [23] can be considered to deal with latter shortages of conventional FMEA.

In general, DEA has been referred to as a fractional mathematical programming technique solely responsible for evaluating the efficiency or performance of Decision-Making Units (DMU) of homogeneous multiple inputs and multiple outputs [24] described Data envelopment analysis (DEA) as a linear programming technique used for determining the relative performance of a set of competing DMUs whenever multiple inputs and outputs make the comparisons cumbersome. It is a non-parametric technique for measuring the technical efficiency of various systems. By technical efficiency, we mean the degree of industry technology level that the production process of a production unit reaches. This can be determined from two perspectives (i) input and (ii) output. From the input aspect under the input condition defined for the system, the technical efficiency is measured by the degree of output maximization and for output perspective under the output condition defined; the technical efficiency is measured by input minimization. In both cases, technical efficiency can be estimated quantitatively as a ratio of output to input.

Hence, DEA has been depicted as a powerful non-parametric performance evaluating technique for calculating the efficiency frontier of homogeneous decision-making units (DMUs) in multi-inputs and multi-response experiments In study of [25] narrated that since its initial proposition a Constant Return to Scale assumption model CCR coined after the first name of the proposers has been used to determine the efficiency of many systems. This model is used for measuring the technical efficiency of individual DMU concerning other DMUs of the same inputs and responses [26]. For the CCR model, the DMUs are evaluated at the optimal production scale. This is the source of the stage of constant returns to scale despite the even scale of production technology is a variable. However, this condition does not hold for real production as many production units are in the constant scale of production. Hence, the technical efficiency of the CCR model comes with characteristics of scale efficiency. In [27], it is reported Variable Return to Scale is based on the assumption BBC as a coinage of FDEA model taking its root from the first letter of its three proposers. This has explained the component of scale efficiency—the possibility of evaluating DMU in the regions of increasing constant and decreasing return to scale. The efficient frontier of the CCR model can be explained either by its primal and dual modes.

The convexity of the production possibility set (PPS) is the primary domain of the DEA models. By implication, every point on the line of the segment between two DMUs belongs to the PPS. Various DEA models have been applied widely in fields—energy, power, water, agriculture, mining, finance, banking, insurance, real estate, transportation, health care, sport, communication, manufacturing, tourism, education, engineering, for product and service-based ventures. However, uncertainty in the data is the bane of the traditional DEA models application. Real-world applications are usually plagued with the menace of uncertainty resulting in shortages. Some DEA models for coping with uncertain have been classified under uncertainty DEA models (UDEA) as depicted according to Pelkani et al., 2020 as shown in Fig. 7.1.

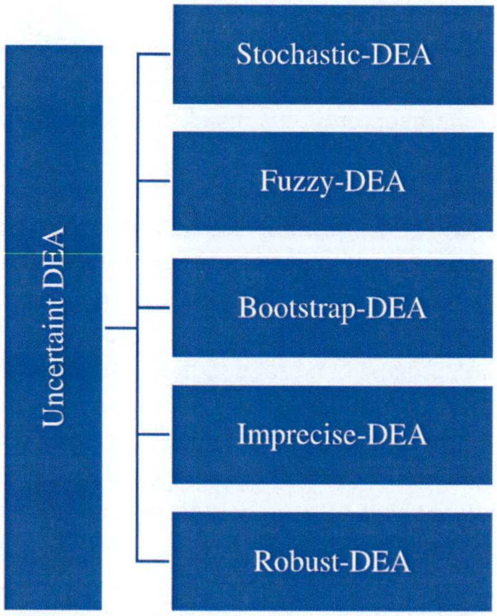

Fig. 7.1 Units of uncertain DEA

In a similar vein, Fuzzy is a logic-based model that dwells on peoples' opinions, observation, and judgments (about imprecise and non-numerical information) for decisions making. They are models that use mathematical formulations to describe and represent vagueness and imprecision in information or premises needed in the decision-making process. Furthermore, Fuzzy sets or models have a high propensity to detect, manipulate, interpret and use data that are marred with uncertainty and vague. Fuzzy logic was first introduced in 1965 by Lotfi Zadeh an Azerbaijani scientist through the proposition of fuzzy set theory. By its classical logic, interpretation can be either true or false with the degree of truth and probabilities comes between 0 and 1. Fuzzy logic thereby uses the degree of truth and probabilities to mathematically modeling the vagueness and ignorance respectively. Fundamentally the application of the truth values could be attributed to some sub-ranges of a continuous variable. For instance, in the gear hobbling various parameters necessary could require several levels for the execution of the gears. Thus, these levels constitute many separate membership functions each deciphering the same and related value to a truth value in the region 0 to 1. This is called fuzzification—a procedure of assigning a numerical value of the input variable of a system to the fuzzy set using the equivalent levels of membership function.

The equivalent of the truth value is then passed as the fuzzy set to evaluate the functionality and control of the gear hobbling. Therefore, it can be noticed that fuzzy set or logic is paving a means for explaining and representing the type and levels of uncertainty associated with the system. The output truth value which is the response variable of the system is not always the same with the fuzzified input truth-value, and it is obtained by a procedure known as defuzzification. Triangular or trapezoidal curves are usually adopted to explicitly define the fuzzy sets. The slope of any of the curves is increasing value with the maximum value of strictly equal to 1 with the length ≥ 0; similarly, the slope can also be of decreasing value. Exclusively, outputs of the fuzzy-based system are borne out of the consensus of the inputs; invariably all the rules can be executed and obeyed when these input values are unavailable or are not trustworthy. This has necessitated the use of weights that can be added to the rule to control the extent of its effects on the output. This optional weighting principle is considered critically according to the prevailing priority, reliability, and consistency of individual rules which can be static or dynamic.

Furthermore, this brief fuzzy set annotation must not be concluded without thinking about how to handle non-numerical values. The expression of certain rules and facts called the linguistic of the fuzzy has been well employed to handle this situation. This provides a means of explaining and representing various linguistic variables in non-numerical terms. We could express variables such as age as being young or old; gender as male and female; geographical areas as being urban or rural, just to mention a few. It is important to emphasize that these parameters need to be expressed with adjectives or adverbs. Fuzzy set theory has been successfully applied to many areas across several countries. Notable of its applications are in the areas of flight aid for helicopters, subway system control, fuel consumption, and precision systems for early recognition of natural disasters.

The Chapter is concern about the application of the integrated fuzzy-based data envelopment analysis. Recently, few researchers have integrated DEA model with fuzzy for enhanced optimum decision making within the FMEA structure, mainly to take adequate corrective actions. These have been demonstrated in a few studies like the efficiency of evaluating risk factors [28], integration of quantitative risk assessment (QRA) and FMEA in [29], and the of DEA comparatively with different weights of risk factors as input and undesirable cost as output [30]. It is presumed that within a short-while, DEA will find its applications within the FMEA domain. Just as DEA has been used to measure the efficiency and productivity of many areas from infrastructure, investments, banks, production systems, it can be used in the same way to support decision making in FMEA procedure.

This chapter is organized to reveal the new approach based on the fuzzy logic and data envelopment analysis introduced are integrated to produce the RPN scores and frontier for taking corrective actions. To understand the concept, an illustration of an aircraft landing problem is made. More so directions for future studies are presented to conclude the chapter.

7.2 Material and Methodology

The framework in [18, 31] illustrates the proposed model and how it prioritized the failure modes (FMs) to obtain corrective actions base on expert judgments for improving the safety performance of the system. The experts expressed their opinions on each risk factor; severity (S), occurrence (O), and detection (D) as specified for FM using defined linguistic terms. An aggregation process where all risk factors are collected using linguistic terms was applied. Subsequently, priority ranking was examined from the riskiest to the least concerning the computed risk priority number ($RPN = S \times O \times D$). Next, corrective actions need two factors: time (T) and cost (C). It is obvious in real life that these factors should be considered for appraising the efficiency of the suggested corrective actions and hence, appropriate safety improvement of the system is determined accordingly. The details of each stage are provided as follows.

7.2.1 Preliminaries of FMEA

In general, the first process for carrying out the FMEA is to develop a well-understanding of what the system is supposed to do when it is operating correctly. In other words, a well understanding of the system process has high importance for going towards the FMEA technique. Therefore, by determining the potential failure modes of the system; three risk factors S, the severity of failure effects; O, the probability of occurrence of failure; D, the probability of failure being detected according to Tables 7.1, 7.2 and 7.3 are evaluated which rated using linguistic terms [18, 32]. As mentioned in the literature, to cope with the ambiguity of FMEA based on expert elicitation, fuzzy logic which is introduced by Zadeh can be used for this purpose [33]. In earlier studies, triangular and trapezoidal fuzzy numbers have been widely used and recognized to be more effective for risk assessment purposes [34–36]. Therefore, both triangular and trapezoidal fuzzy numbers were utilized to map linguistic

Table 7.1 Severity evaluation element

Typical FMEA			Fuzzy scale	
Rank	Effect	Criteria	Linguistic terms	Corresponding fuzzy number
9, 10	Very high	According to this failure mode, death or mission losses occur	Very high	(0.8, 0.9, 1)
7, 8	High	Severe injury, damage, or incompleteness on a mission may occur by the effect of failure mode	High	(0.5, 0.6, 0.7, 0.8)
4, 5, 6	Moderate	Loss of function by system or component may occur	Medium	(0.4, 0.5, 0.6)
2, 3	Low	Minor injury or depreciation in system performance	Low	(0.2, 0.3, 0.4, 0.5)
1	Minor	Usually, injury or depreciation in system performance does not occur	Very low	(0, 0.1, 0.2)

Table 7.2 Probability of occurrence evaluation element

Typical FMEA				Fuzzy scale	
Rank	Likelihood	Criteria	Predicted failure frequency	Linguistic terms	Corresponding fuzzy number
10	Very high	Failure is almost unavoidable	> 1 in 2	Very high	(0.8, 0.9, 1)
9			1 in 8		
8	High	Repeated failures	1 in 20	High	(0.5, 0.6, 0.7, 0.8)
7			1 in 40		
6	Moderate	Occasional failures	1 in 80	Medium	(0.4, 0.5, 0.6)
5			1 in 400		
4			1 in 1000		
3	Low	Rare failures	1 in 4000	Low	(0.2, 0.3, 0.4, 0.5)
2			1 in 20,000		
1	Remote	Failure is unlikely	< 1 in 10^6	Very low	(0, 0.1, 0.2)

terms to the fuzzy membership function. The reason for using these two types of fuzzy numbers is that under some weak assumptions, the defined membership functions can directly meet the appropriate optimization criteria [37]. The membership function can be defined both trapezoidal and triangular fuzzy numbers as follows [38]:

(a) In triangular form, $\tilde{A} = (a_1, a_2, a_3)$

$$V_i(x) = \left\{ \begin{array}{ll} 0, & x < a_1 \\ \frac{x-a_1}{a_2-a_1}, & a_1 \leq x \leq a_2 \\ \frac{a_3-x}{a_3-a_2}, & a_2 \leq x \leq a_3 \\ 0, & x > a_3 \end{array} \right\} \tag{7.1}$$

(b) In trapezoidal form, $\tilde{A} = (a_1, a_2, a_3, a_4)$

$$V_i(x) = \left\{ \begin{array}{ll} 0, & x < a_1 \\ \frac{x-a_1}{a_2-a_1}, & a_1 \leq x \leq a_2 \\ 1, & a_2 \leq x \leq a_3 \\ \frac{a_4-x}{a_4-a_3}, & a_3 \leq x \leq a_4 \\ 0, & x > a_4 \end{array} \right\} \tag{7.2}$$

Table 7.3 Detectability evaluation element

Typical FMEA			Fuzzy Scale	
Rank	Detectability	Criteria	Linguistic terms	Corresponding Fuzzy number
10	Non-detection	Program control or any user cannot detect a potential cause or failure mode	None	(0, 0.1, 0.2)
9	Very low	Program control or user will not detect a potential cause or failure mode	Very low	(0.1, 0.2, 0.3)
7, 8	Low	Program control or user is not likely to detect a potential cause or failure mode	Low	(0.2, 0.3, 0.4, 0.5)
5, 6	Moderate	Program control or user may detect a potential cause or failure mode	Medium	(0.4, 0.5, 0.6)
3, 4	Fairly high	Program control or user has a proper chance of detecting a potential cause or failure mode	Fairly high	(0.5, 0.6, 0.7, 0.8)
2	High	Program control or user will almost certainly detect a potential cause or failure mode	High	(0.7, 0.8, 0.9)
1	Very high	Program control or user will certainly detect a potential cause or failure mode	Very high	(0.8, 0.9, 1)

7.2.2 Expert Judgment

To obtain much more reliable results as the output of any risk assessment methods, it is better to use multi-expert opinions as an independent resource. However, in the real case, each expert has a different weight according to his or her own background such as job tenure, education, experience, and of course paradigm shift. Consequently, as discussed in [21, 39] employing a heterogeneous group of expert have enough superiority for precision compare to homogenous one. Thus, in this study standard method is engaged for weighting experts group selected in a heterogeneous manner.

7.2.2.1 Expert Weighing

There are many methods of assigning specific weight to an expert in an employed group. A simple averaging method having criteria such as age, personal experience, job tenure, and education level have been considered in the literature for expert weighting. However, this method does not show high effectiveness [39–44]. Therefore, to assign a much more reliable weight to employed experts, an extension of the fuzzy analytic hierarchy process (FAHP) was introduced to handle this issue. The conventional AHP is a well-known method commonly used in multi-criteria decision-making problems. However, it cannot deal with subjective knowledge and as a result, FAHP has been developed to solve the AHP problems [45]. In other words, the main purpose of AHP is collecting expert opinions but it cannot take into cognizance human thinking ability. Several FAHP techniques have been proposed in the past decades with the two most important ones introduced by Buckley (1985) and Chang (1996) which uses trapezoidal and triangular fuzzy membership function respectively for pairwise comparison scale [46, 47]. In this study, an extension of Buckley's method is used for weighting the experts due to the limitations of other techniques. Furthermore, it should be noted that for Buckley's method, illogical zero weight may also be obtained for the selected criteria (for more details see [48, 49]). In pursuant to the adoption of the aforementioned methods, experts' weight is computed more reliably based on their knowledge and backgrounds. Therefore, the computed weights are vital and representative of the relative superiority of the employed experts.

So far, the weight of each expert is computed more reliably based on their knowledge and background. Therefore, the computed weights are vital to represent the relative superiority of the employed experts.

7.2.2.2 Aggregation Procedure

It is vital to aggregate expert judgments to reach a concluded deal because each expert may have a variety of opinions according to her/his knowledge and experience in the pertinent scope. Therefore, using an aggregation procedure has high importance. Many methods available in the literature include max–min Delphi method, similarity aggregation method (SAM), α-cut, arithmetic averaging operation, etc. However, there is no evidence to show which method has enough superiority to compare with the other one [50]. As mentioned earlier, experts generally prefer to express their opinion in qualitative terms. In this regard, many ambiguity and uncertainty may arise during the risk assessment. Fuzzy set theory as an alternative and proper tool can deal with this situation to minimize the uncertainty in the risk assessment procedure. Therefore, introducing an approach based on fuzzy set theory to transfer linguistic expert opinions into fuzzy set numbers to aggregate their opinions into a crisp value is required. The aggregation procedure of expert judgment in the fuzzy logic system is divided into three steps as itemized below.

Step 1: Obtaining linguistic terms of each risk factor (S, O, and D) based on expert judgment
Step 2: Converting linguistic terms into fuzzy corresponding numbers
Step 3: Converting fuzzy numbers into fuzzy set scores (FSS).

Step 1: The purpose of step 1 is to collect the risk factors of each FM based on linguistic terms expressed by the concerned industry experts. In this regard, several experts were consulted, and they were sent a questionnaire (by email) to judge the risk factors of each FM. In this way, experts are more relaxed to express their opinion in linguistic terms compared to the numerical scale [51]. The linguistic terms to represent the score of risk factors concerning FMs were provided in different categories (see Tables 7.1, 7.2 and 7.3). These categories are selected based on Saaty's study which posited that the proper number for expert judgment at a specific time is between five and nine. In other words, the common capacity of the human memory is seven plus-minus two chunks [52].

Step 2: The sum-production method, which is used in this study for the aggregation process is represented as follows.

$$Z_i = \sum_{j=1}^{n} w_j \otimes f_{ij}, \quad i = 1, 2, \ldots, m \quad j = 1, 2, \ldots, n \tag{7.3}$$

where Z_{ir} denotes the aggregated fuzzy number for risk factor i for each FM, w_j represents the weight of experts j, and f_{ij} is corresponding fuzzy number of FM r given by expert j. n and m are the numbers of experts and FMs respectively.

Step 3: FSS denotes a crisp value that is based on experts' opinion aggregation for any risk factors. To defuzzify a quantifiable outcome in the fuzzy set theory, many common techniques are available. Among them, the center of Area (CoA) is more popular. The details of the computation for the defuzzification procedure for both techniques are provided as follows.

CoA mathematically represents in Eq. 7.4.

$$X^* = \frac{\int \upsilon_i(x)x dx}{\int \upsilon_i(x) dx} \tag{7.4}$$

were

X^*	Defuzzified output.
$\upsilon_i(x)$	Aggregated membership function.
x	output variable.

Equation 7.4 can be engaged to both trapezoidal and triangular fuzzy numbers and it is shown as:

Defuzzification of triangular fuzzy number $\tilde{A} = (a_1, a_2, a_3)$ is:

$$X^* = \frac{\int_{a_1}^{a_2} \frac{x-a_2}{a_2-a_1} x dx + \int_{a_2}^{a_3} \frac{a_3-x}{a_3-a_2} x dx}{\int_{a_1}^{a_2} \frac{x-a_2}{a_2-a_1} dx + \int_{a_2}^{a_3} \frac{a_3-x}{a_3-a_2} dx} = \frac{1}{3}(a_1 + a_2 + a_3) \tag{7.5}$$

Defuzzification of trapezoidal fuzzy number $\tilde{A} = (a_1, a_2, a_3, a_4)$ can be gained by Eq. 7.6

$$X^* = \frac{\int_{a_1}^{a_2} \frac{x-a_1}{a_2-a_1} x dx + \int_{a_2}^{a_3} x dx + \int_{a_3}^{a_4} \frac{a_4-x}{a_4-a_3} x dx}{\int_{a_1}^{a_2} \frac{x-a_1}{a_2-a_1} dx + \int_{a_2}^{a_3} dx + \int_{a_3}^{a_4} \frac{a_4-x}{a_4-a_3} dx}$$
$$= \frac{1}{3} \frac{(a_4 + a_3)^2 - a_4 a_3 - (a_1 + a_2)^2 + a_1 a_2}{(a_4 + a_3 - a_1 - a_2)} \tag{7.6}$$

So far, the crisp value of all three risk factors (RPN) is computed. In FMEA technique, RPN has been used as a divisor of FMs into the degree of priority for taking corrective action in order to reduce the risk and accordingly improve the safety performance of the system. However, considering the time and cost implication of the necessary corrective actions brings many limitations for decision makers. In the next section, Date envelopment analysis (DEA) is described to show how decision makers can effectively prioritize FMs for taking effective corrective actions base on the RPN, the time required, and cost incurred as FMs performance indicators.

7.2.3 *Date Envelopment Analysis (DEA)*

To improve the safety performance of the system, many corrective actions are recommended to reduce the value of S, O, and D. However, system management of the company prefers to make a decision based on less time and cost. For this purpose, the ideal and effective situation for making a decision is to consider a higher score of RPN concerning the minimum cost incurred and time taken. In other words, optimum corrective actions can be determined for the highest risky FMs in a short period and at a minimal cost. Applying corrective actions should be geared toward minimizing the risk of a system to be at least tolerable limit. These corrective actions can be divided into three categories including inherent safety, engineering safety, and procedural safety. These three categories have been discussed extensively by [53, 54]. Holistically, all the FMs are considered to be effectively prioritized. In this regard, DEA model can handle this situation which is described as follows.

7.2.3.1 DEA Methodology

Data Envelopment Analysis was carried out to determine the efficiency of the DMUs at input and output orientations. In DEA the efficiency of a Decision-Making Unit (DMU) is computed based on its inputs consumption and outputs production. Between homogeneous DMUs, a DMU is efficient if it can produce more outputs by consuming fewer inputs. Mathematically this means that the fraction of weighted outputs average over the average of the weighted inputs should maximize such that this fraction for other DMUS remains in [0, 1] interval. As applied in [55], if y_{ro} is the *r*th output and x_{ij} is the *i*th input for *j*th DMU, say DMU_j j = 1,…,n, and u_r and v_i, consider the output and input weights for r = 1,…,s and i = 1,…m, respectively, the CCR input-orientation model is as follows where denote DMUunder evaluation DMU;

$$\begin{aligned}
\text{Max} \quad & \sum_{r=1}^{s} u_r y_{ro} \\
St. \quad & \sum_{i=1}^{m} v_i x_{io} = 1 \\
& \sum_{r=1}^{s} u_r y_{rj} - \sum_{i=1}^{m} v_i x_{ij} \leq 0 \quad j = 1, \ldots, n \\
& u_r \geq 0 \quad r = 1, \ldots, s \\
& v_i \geq 0 \quad i = 1, \ldots, m
\end{aligned} \tag{7.7}$$

Here each potential FMEA of the system is seen as a DMU, inputs are the potential failure modes S, O, D, and one over RPN obtained after defuzzification, outputs are the inverse of the FMs performance indicators cost (C) and time (T) requires for taking corrective actions. The objective function gives the efficiency of each FMEA.

To ensure that the efficiencies gotten are the real benchmark and they yield better FMEA relatively, the above primal model was converted to dual model. This was the view of appraising through the input and output slacks, the impact of relatively efficient (best) FMs to the underperforming FMs. The CCR dual model is given as

$$\begin{aligned}
\text{Min} \quad & \theta \\
St. \quad & \sum_{j=1}^{n} \lambda_j x_{ij} \leq \theta x_{io} \quad i = 1, \ldots, m \\
& \sum_{j=1}^{n} \lambda_j y_{rj} \leq y_{ro} \quad r = 1, \ldots, s \\
& \lambda_j \geq 0 \quad j = 1, \ldots, n
\end{aligned} \tag{7.8}$$

Efficient FMEAs are those with lower S, O, D, and RPN risk factors and moderately lower time and cost requirements for taking corrective actions. Conversely, inefficient FMEAs are those with higher S, O, D, and RPN, higher time and cost for taking corrective actions. PIM DEA version 3.2 was used to solve the model.

According to the developed smart fuzzy FMEA framework, decision makers have enough capability to recognize which corrective action is efficient to FMEA system under evaluation. To be thoroughly acquainted with the proposed approach, in the next section a real case study is considered purposeful.

7.3 Application of Study

To explain the proposed approach clearly, the detail of one of the numeric failure modes (The simplest aircraft landing systems) is chosen as a case of study, because in recent years many accidents related to landing systems have been reported by airlines.

As shown in system landing process in [18, 31], the simplest aircraft landing system contains five functions which are raising gear, coming down gear, automatically test, fault reporting system, and record the result of automatic test. The FMs studied are provided in Table 7.4 (for more details see [18, 31]).

As it mentioned earlier, to expedite an unbiased judgment for each element of FMEA, a heterogeneous group of experts is employed within the categories listed below.

Expert 1: Safety engineer who has been working as a risk assessor and safety auditor in varieties of the industry for more than four years with master qualification.
Expert 2: Safety engineer who has been working as a risk assessor and safety auditor in varieties of the industry for more than 15 years with a master qualification.
Expert 3: Academic staff in an aerospace engineering discipline who has been working as consultant and trainer for aerospace industries for more than 10 years with Ph.D. qualification.
Expert 4: Engineers with experience in reliability analysis who has been working at reliability center for more than 20 years with BS qualification.

Table 7.4 FMEA of the aircraft landing system

Process Step		Potential failure modes
FM1	Raising gear	Fault in raising the wheels
FM2		Raising wheels earlier than the specified time
FM3	Coming down gear	Fault in coming down the wheels
FM4		Coming down wheels earlier than the specified time
FM5	Automatically test	Fault on the run
FM6		Wrong run
FM7	Fault reporting system	Fault on the run
FM8		Wrong run
FM9	Record the result of automatically test	Fault on the run
FM10		Wrong run

The fuzzy AHP method was used to compute each expert's capability and assigning the respective weights. In this regard, the following weights 0.14, 0.18, 0.33, and 0.35 are given to E1, E2, E3, and E4 respectively.

As an illustration that FM8 Fault in raising the wheels; with respect to the definition of fuzzy numbers provided in Tables 7.1, 7.2 and 7.3, the computation of element O is done as follows. The qualitative terms, given by four experts, fall into "M", "M", "M", and "H" categories. The integrated fuzzy number is attained as follows:

$$\begin{aligned} \mathrm{Z_{FM8}} &= (\mathrm{w_{E1}} + \mathrm{w_{E2}} + \mathrm{w_{E3}}) \times \mathrm{f_M(x)} + \mathrm{w_{E4}} \times \mathrm{f_H(x)} \\ &= (0.14 + 0.18 + 0.33) \times (0.4, 0.5, 0.5, 0.6) + 0.35 \times (0.7, 0.8, 0.9) \\ &= (0.505, 0.505, 0.605, 0.705) \end{aligned}$$

Equations 7.4–7.6 are employed to Defuzzifier of each failure mode. The computation of FM8 for element O is done as an example and the results of other failure modes to having crisp values with three decimal places are provided as the following:

$$\begin{aligned} X^* &= \frac{1}{3}\frac{(0.705 + 0.605)^2 - (0.705 \times 0.605) - (0.505 + 0.505)^2 + (0.505 \times 0.505)}{(0.705 + 0.605 - 0.505 - 0.505)} \\ &= 0.6 \end{aligned}$$

The last step of defuzzification is multiplying X^* to 10:

$$X^* \times 10 = 5.87.$$

According to RPN scores and their priority, corrective actions based on the expert specialist of the system should be recommended. As it mentioned, modifying corrective actions needs time and cost. Therefore, DEA can help decision makers to prioritize all failure modes efficiently. In this regard, it is asked experts to provide information in the terms of time and cost for modifying corrective actions. Accordingly, DEA is applied to understand which FMs have enough efficiency for taking to account.

FM-DMU 3, 4, 7, 9, and 10 are efficient, they thereby possess lower cost and time requirements for taking corrective actions. This also denotes that those efficient FMs will be highly prioritized for taking corrective measures due to lower cost and time indicators. DMUs 1, 2, 5, 6, and 8 are inefficient though they possess higher cost and time are required for taking corrective actions and they are least prioritized. Applying conventional FMEA ranking and efficiency scores conjunctively, then the order of priority for taking corrective actions is given as shown below where FMEA 9 needed to be prioritized than 10, 7, 4 and 3 to be the least;

$$\mathrm{FM9} > \mathrm{FM10} > \mathrm{FM7} > \mathrm{FM4} > \mathrm{FM3}$$

Interestingly by factoring in time and cost, the above order conformed adequately well with the assertion of the conventional FMEA ranking that FM with the highest RPN is to the given the highest priority. Similarly, for the inefficient FMs, the order of priority for taking corrective actions based on lower performance indicators according to both RPN and efficiency scores (the higher the efficiency scores the better for taking corrective actions for the FM) has shown that FM1 must to be prioritized than 8, 2, 5 and 6 to be the least;

$$\text{FM1} > \text{FM8} > \text{FM2} > \text{FM5} > \text{FM6}$$

Here, the conventional FMEA does not hold for the inefficient FMEAs after time and cost have been considered. By combining both outcomes, the ranking of the FMEAs as obtained by FSFMEA can be slated as.

$$\text{FM9} > \text{FM10} > \text{FM7} > \text{FM4} > \text{FM3} > \text{FM1} > \text{FM8} > \text{FM2} > \text{FM5} > \text{FM6}$$

It is necessary to clarify that inefficient frontier only identified FMs that cannot yield the indicative response to the conditions for initiating corrective actions and not that they are not important to the overall safety of the system. Instead, it is expected that safety personnel should take corrective actions on the efficient FMs and more proactive measures on the inefficient FMs especially FM6.

Furthermore computation shows that the safety performance of the aircraft landing system understudied shows that time T with a total weight of 14.97 and cost C with a total weight of 15.32 are germane to the technical efficiency and are very pivotal in the assessment of FMs rather than using only the risk factor RPN. Similarly, from the input weights, occurrence (O) with the highest total weight of 10.95 revealed that the degree of occurrence of these safety indicators is high. Detection (D) value being infinitesimally small connotes that it does not have an effect on the efficiency status and is not pertinent for corrective actions to be initiated for the mentioned efficient FMs. But this could further imply a signal to the safety engineer or manager to be more proactive rather than being reactive especially for the inefficient FMEA 1, 2, 5. 6, and 8. The RPN of total weight 4.37 makes it relative to the decision-making process and has been supported by the output of DEA.

Worthy of mentioning is the smartness of the framework that identified FMs 9 and 10 as a typical case. They were ranked 2nd and 4th positions by conventional RPN meaning that FM9 more is riskier than FM10. However, the two FMs have the same priority despite that the RPN of FM9 is higher than FM10. It is interesting to note that both have the same cost implications of 10.525 with FM9 consuming time of 2.14 more than FM10 that consumed the time of 1.84. By FSFMEA, FM10 should have more priority than FM9, but relatively to the conventional FMEA both should be prioritized equally or. Safety assessors could rationally give priority to FM10 before FM9. Additionally, conventional FMEA ranked FM6 highest which implied that it is to be highly prioritized; conversely, smart FMEA revealed that it must be of the least priority. This can be explained by looking at the priority status of the highly prioritized FMs in the perspective of both conventional FMEA and the newly proposed FSFMEA. The total RPN of FMs 9 and 10 put at 401.490 is higher than RPN value of 243.810 for FM6. Furthermore, the total time for FM9 and 10 of 3.98 and total cost of 21.05 are lower than the time of 24.25 and the cost of 46.88

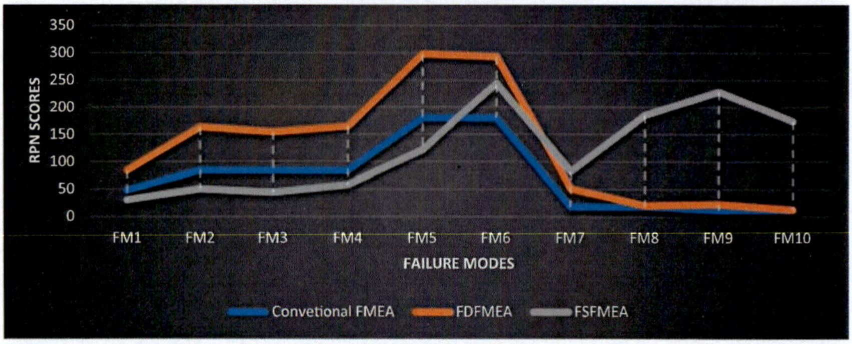

Fig. 7.2 The comparison of different FMEA approaches considering RPN scores

required for taking corrective action for FM6. From this analysis, the cost and time requirements for taking corrective action for FM6 alone is more than twice what is required for both FMs 9 and 10 put together even when the risk level of FMs 9 and 10 occurring twice is more than the level of a single occurrence of FM6. It is therefore reasonable on the part of the safety engineer or officer to give priority to taking corrective actions for FMs 9 and 10 as specified by the proposed FSFMEA rather than for FM6 as obtained by conventional FMEA. This is another insightful advantage of FSFMEA over the conventional FMEA.

In addition, as it can be seen from Fig. 7.2, a comparison is done to compare the results of RPN scores in different approaches including conventional FMEA [31], fuzzy developed FMEA (FDFMEA) [18], and fuzzy smart FMEA (FSFMEA). According to the RPN scores, FDFMEA and FSFMEA demonstrated the same trend. However, in some modes like FM8, FM9, and FM10; it is found to be different considerably. Moreover, based on FMEA rules, the priority for decreasing risk amount is placed on the highest-ranked FMs which are approximately the same in FDFMEA and FSFMEA approaches. Besides, both approaches show that the new RPN scores can handle the shortages of conventional RPNs. This has been buttressed as mentioned in the literature. Furthermore, such a risk assessment study is based on experts' opinions and noting that many things can influence their opinions. In this regard, FSFMEA has more advantages over FDFMEA, because fewer computations are required to aggregate experts' opinions and in FDFMEA attending to the value of the relaxation factor has high necessity whereas in FSFMEA approach it is not.

7.4 Conclusion

FMEA has been well used as an efficient technique to help assessors to identify, prioritize, and prevent potential FMs in the design or process of an industry. However, conventional FMEA has been criticized because of many reasons such as uncertainty and vagueness in the efficiency of corrective actions; accordingly, many approaches are represented to cope with shortages of conventional FMEA. This paper presented a modification of Yazdi et al.'s approach [18] to combine the risk evaluations by employing multi-experts' opinions using fuzzy set theory. The modified approach adds a DEA process to recognize the efficiency of recommended corrective actions. An example and application results have illustrated that the modified approach in this study has the advantage over Yazdi et al.'s approach and can also provide much more information to make better decisions to decrease the risk level and subsequently improve the safety performance of the system. However, a limitation of the study arises to authors, and it is the proposed model cannot guarantee that the highest risky FM will be reduced according to this model. In other words, it has a high necessity to provide corrective actions for the highest RPN either be efficient or not. Obviously, in some cases, the consequence(s) of a specific FM can be much more than the cost and time of the corrective actions. As a direction for further studies, the authors would like to apply sensitivity analysis in the proposed approach with consideration of relative importance weights of risk factors. In addition, subjectivity and objectivity uncertainties can be integrated to get more realistic results.

References

1. H.-C. Liu, L. Liu, N. Liu, Risk evaluation approaches in failure mode and effects analysis: a literature review. Expert Syst. Appl. **40**, 828–838 (2013). https://doi.org/10.1016/j.eswa.2012.08.010
2. H.-C. Liu, L. Liu, Q.-H. Bian, Q.-L. Lin, N. Dong, P.-C. Xu, Failure mode and effects analysis using fuzzy evidential reasoning approach and grey theory. Expert Syst. Appl. **38**, 4403–4415 (2011). https://doi.org/10.1016/j.eswa.2010.09.110
3. K.H. Chang, Generalized multi-attribute failure mode analysis. Neurocomputing **175**, 90–100 (2015). https://doi.org/10.1016/j.neucom.2015.10.039
4. B. Joo, S. Kim, S. Kim, Y.H. Moon, FMEA for the reliability of hydroformed flanged part for automotive application. J. Mech. Sci. Technol. **27**, 63–67 (2013). https://doi.org/10.1007/s12206-012-1226-5
5. K.-H. Chang, C.-H. Cheng, A risk assessment methodology using intuitionistic fuzzy set in FMEA. Int. J. Syst. Sci. **41**, 1457–1471 (2010). https://doi.org/10.1080/00207720903353633
6. M.D. Dargahi, S. Naderi, S.A. Hashemi, M. Aghaiepour, Z. Nouri, S.K. Sahneh, Use FMEA method for environmental risk assessment in ore complex on wildlife habitats. Hum. Ecol. Risk Assess. **22**, 1123–1132 (2016). https://doi.org/10.1080/10807039.2015.1106912
7. D.H. Stamatis, Failure Mode and Effect Analysis: FMEA from Theory to Execution (2003). https://doi.org/10.2307/1268911
8. J.K. Chen, Utility priority number evaluation for FMEA. J. Fail. Anal. Prev. **7**, 321–328 (2007). https://doi.org/10.1007/s11668-007-9060-2
9. C.S. Carlson, *Effective FMEAs Achieving Safe, Reliable, and Economical Products and Processes Using Failure Mode and Effects Analysis* (Wiley, 2012)
10. D. Zhou, Y. Tang, W. Jiang, A modified model of failure mode and effects analysis based on generalized evidence theory. Math. Probl. Eng. **2016** (2016). https://doi.org/10.1155/2016/4512383
11. W. Song, X. Ming, Z. Wu, B. Zhu, Failure modes and effects analysis using integrated weight-based fuzzy TOPSIS. Int. J. Comput. Integr. Manuf. **26**, 1172–1186 (2013). https://doi.org/10.1080/0951192X.2013.785027

12. E. Bozdag, U. Asan, A. Soyer, S. Serdarasan, Risk prioritization in failure mode and effects analysis using interval type-2 fuzzy sets. Expert Syst. Appl. **42**, 4000–4015 (2015). https://doi.org/10.1016/j.eswa.2015.01.015
13. Y. Zhang, J. Andrews, S. Reed, M. Karlberg, Maintenance processes modelling and optimisation. Reliab. Eng. Syst. Saf., 1–11 (2017). https://doi.org/10.1016/j.ress.2017.02.011
14. H.-C. Liu, J.-X. You, X.-J. Fan, Q.-L. Lin, Failure mode and effects analysis using D numbers and grey relational projection method. Expert Syst. Appl. **41**, 4670–4679 (2014). https://doi.org/10.1016/j.eswa.2014.01.031
15. H. Li, J.Y. Guo, M. Yazdi et al., Supportive emergency decision-making model towards sustainable development with fuzzy expert system. Neural Comput. Appl. **33**, 15619–15637 (2021). https://doi.org/10.1007/s00521-021-06183-4
16. H. Liu, FMEA Using Uncertainty Theories and MCDM Methods (2016). https://doi.org/10.1007/978-981-10-1466-6
17. C. Dağsuyu, E. Göçmen, M. Narlı, A. Kokangül, Classical and fuzzy FMEA risk analysis in a sterilization unit. Comput. Ind. Eng. **101**, 286–294 (2016). https://doi.org/10.1016/j.cie.2016.09.015
18. M. Yazdi, S. Daneshvar, H. Setareh, An extension to fuzzy developed failure mode and effects analysis (FDFMEA) application for aircraft landing system. Saf. Sci. **98**, 113–123 (2017). https://doi.org/10.1016/j.ssci.2017.06.009
19. T.A. Montgomery, D.R. Pugh, S.T. Leedham, S.R. Twitchett, FMEA automation for the complete design process, in *Proceedings of 1996 Annual Reliability and Maintainability Symposium*, pp. 30–36 (1996). https://doi.org/10.1109/RAMS.1996.500638
20. S. Daneshvar, M. Yazdi, K.A. Adesina, Fuzzy smart failure modes and effects analysis to improve safety performance of system: Case study of an aircraft landing system. Qual. Reliab. Engng. Int. **36**, 890–909 (2020). https://doi.org/10.1002/qre.2607
21. S. Helvacioglu, E. Ozen, Fuzzy based failure modes and effect analysis for yacht system design. Ocean Eng. **79**, 131–141 (2014). https://doi.org/10.1016/j.oceaneng.2013.12.015
22. K.C. Chai, C.H. Jong, K.M. Tay, C.P. Lim, A perceptual computing-based method to prioritize failure modes in failure mode and effect analysis and its application to edible bird nest farming. Appl. Soft Comput. J. **49**, 734–747 (2016). https://doi.org/10.1016/j.asoc.2016.08.043
23. A. Charnes, W.W. Cooper, E. Rhodes, Measuring the efficiency of decision making units. Eur. J. Oper. Res. **2**, 429–444 (1978). https://doi.org/10.1016/0377-2217(78)90138-8
24. L.C.S. Rocha, A.P. Paiva, E.J. Paiva, P.P. Balestrassi, Comparing DEA and principal component analysis in the multiobjective optimization of P-GMAW process. J. Brazilian Soc. Mech. Sci. Eng. **38**, 2513–2526 (2016). https://doi.org/10.1007/s40430-015-0355-z
25. J. Wu, L. Liang, Y. Chen, DEA game cross-efficiency approach to Olympic rankings. Omega **37**, 909–918 (2009). https://doi.org/10.1016/j.omega.2008.07.001
26. A. Al-Refaie, Optimizing performance with multiple responses using cross-evaluation and aggressive formulation in data envelopment analysis. IIE Trans. (Institute Ind. Eng.) **44**, 262–276 (2012). https://doi.org/10.1080/0740817X.2011.566908
27. R. Ma, L. Yao, M. Jin, P. Ren, The DEA game cross-efficiency model for supplier selection problem under competition. Appl. Math. Inf. Sci. **8**, 811–818 (2014). https://doi.org/10.12785/amis/080242
28. A.U. Baloch, H. Mohammadian, Fuzzy failure modes and effects analysis by using fuzzy Vikor and Data Envelopment Analysis-based fuzzy AHP. Int. J. Adv. Appl. Sci. **3** (2016)
29. A. Kengpol, S. Tuammee, The development of a decision support framework for a quantitative risk assessment in multimodal green logistics: an empirical study. Int. J. Prod. Res. **54**, 1020–1038 (2016). https://doi.org/10.1080/00207543.2015.1041570
30. M. Jahangoshai Rezaee, A. Salimi, S. Yousefi, Identifying and managing failures in stone processing industry using cost-based FMEA. Int. J. Adv. Manuf. Technol. **88**, 3329–3342 (2017). https://doi.org/10.1007/s00170-016-9019-0

31. C.A. Ericson, *Hazard Analysis Techniques for System Safety* (Wiley, Hoboken, NJ, USA, 2005). https://doi.org/10.1002/0471739421
32. M. Yazdi, O. Korhan, S. Daneshvar, Application of fuzzy fault tree analysis based on modified fuzzy AHP and fuzzy TOPSIS for fire and explosion in the process industry, Int. J. Occupational Saf. Ergonomics, **26**(2), 319-335, (2020). https://doi.org/10.1080/10803548.2018.1454636
33. L. Zadeh, Fuzzy Sets. Inf. Control. **8**, 338–353 (1965)
34. R. Ferdous, F. Khan, R. Sadiq, P. Amyotte, B. Veitch, Analyzing system safety and risks under uncertainty using a bow-tie diagram: an innovative approach. Process Saf. Environ. Prot. **91**, 1–18 (2013). https://doi.org/10.1016/j.psep.2011.08.010
35. M. Yazdi, F. Khan, R. Abbassi, R. Rusli, Improved DEMATEL methodology for effective safety management decision-making. Saf. Sci. **127**, 104705 (2020). https://doi.org/10.1016/j.ssci.2020.104705
36. M. Yazdi, S. Kabir, A Fuzzy Bayesian Network approach for risk analysis in process industries. Process Saf. Environ. Prot. (2017). https://doi.org/10.1016/j.psep.2017.08.015
37. W. Pedrycz, Why triangular membership functions? Fuzzy Sets Syst. **64**, 21–30 (1994). https://doi.org/10.1016/0165-0114(94)90003-5
38. T.J. Ross, Fuzzy Logic with Engineering Applications (2009). https://doi.org/10.1002/9781119994374
39. M. Yazdi, F. Nikfar, M. Nasrabadi, Failure probability analysis by employing fuzzy fault tree analysis. Int. J. Syst. Assur. Eng. Manag., 1–17 (2017). https://doi.org/10.1007/s13198-017-0583-y
40. N. Ramzali, M.R.M. Lavasani, J. Ghodousi, Safety barriers analysis of offshore drilling system by employing Fuzzy event tree analysis. Saf. Sci. **78**, 49–59 (2015). https://doi.org/10.1016/j.ssci.2015.04.004
41. M.R. Miri Lavasani, J. Wang, Z. Yang, J. Finlay, Application of fuzzy fault tree analysis on oil and gas offshore pipelines. Int. J. Mar. Sci. Eng. **1**, 29–42 (2011). http://www.sid.ir/en/VEWSSID/J_pdf/1035520110104.pdf
42. F. Yan, K. Xu, X. Yao, Y. Li, Fuzzy Bayesian network-bow-tie analysis of gas leakage during biomass gasification. PLoS ONE **11**, e0160045 (2016). https://doi.org/10.1371/journal.pone.0160045
43. S.M. Lavasani, N. Ramzali, F. Sabzalipour, E. Akyuz, Utilisation of Fuzzy Fault Tree Analysis (FFTA) for quantified risk analysis of leakage in abandoned oil and natural-gas wells. Ocean Eng. **108**, 729–737 (2015). https://doi.org/10.1016/j.oceaneng.2015.09.008
44. S.M. Lavasani, A. Zendegani, M. Celik, An extension to Fuzzy Fault Tree Analysis (FFTA) application in petrochemical process industry. Process Saf. Environ. Prot. **93**, 75–88 (2015). https://doi.org/10.1016/j.psep.2014.05.001
45. M. Gul, A.F. Guneri, A fuzzy multi criteria risk assessment based on decision matrix technique: a case study for aluminum industry. J. Loss Prev. Process Ind. **40**, 89–100 (2016). https://doi.org/10.1016/j.jlp.2015.11.023
46. J.J. Buckley, Fuzzy hierarchical analysis. Fuzzy Sets Syst. **17**, 233–247 (1985). https://doi.org/10.1016/0165-0114(85)90090-9
47. D.-Y. Chang, Applications of the extent analysis method on fuzzy AHP. Eur. J. Oper. Res. **95**, 649–655 (1996). https://doi.org/10.1016/0377-2217(95)00300-2
48. M. Yazdi, Hybrid probabilistic risk assessment using fuzzy FTA and fuzzy AHP in a process industry. J. Fail. Anal. Prev. **17**, 756–764 (2017). https://doi.org/10.1007/s11668-017-0305-4
49. H.K. Chan, X. Wang, Fuzzy extent analysis for food risk assessment, in *Fuzzy Hierarchical Model Risk Assess* (Springer London, London, 2013), pp. 89–114. https://doi.org/10.1007/978-1-4471-5043-5_6
50. Y. Liu, Z.P. Fan, Y. Yuan, H. Li, A FTA-based method for risk decision-making in emergency response. Comput. Oper. Res. **42**, 49–57 (2014). https://doi.org/10.1016/j.cor.2012.08.015
51. B.M. Ayyub, G.J. Klir, Uncertainty Modeling and Analysis in Engineering and the Sciences (2006). https://doi.org/10.1201/9781420011456

52. T.L. Saaty, M.S. Ozdemir, Why the magic number seven plus or minus two. Math. Comput. Model. **38**, 233–244 (2003). https://doi.org/10.1016/S0895-7177(03)90083-5
53. M. Yazdi, Ignorance-aware safety and reliability analysis: A heuristic approach. Qual. Reliab. Eng. Int. **36**, 652–674 (2020). https://doi.org/10.1002/qre.2597
54. M. Yazdi, A perceptual computing–based method to prioritize intervention actions in the probabilistic risk assessment techniques. Qual. Reliab. Eng. Int. **36**, 187–213 (2020). https://doi.org/10.1002/qre.2566
55. S. Daneshvar, G. Izbirak, A. Javadi, Sensitivity analysis on modified variable returns to scale model in Data Envelopment Analysis using facet analysis. Comput. Ind. Eng. **76**, 32–39 (2014). https://doi.org/10.1016/j.cie.2014.07.016

Chapter 8
Introducing a Probabilistic-Based Hybrid Model (Fuzzy-BWM-Bayesian Network) to Assess the Quality Index of a Medical Service

Mohammad Yazdi, Sidum Adumene, and Esmaeil Zarei

8.1 Introduction

To assess the quality of hospital services, health care systems, and similar application domains, multi-criteria decision-making (MCDM) is typically utilized. In the state of arts, extensive researches have been used applying MCDM tools in health care systems [1–6]. In MCDM tools, a set of alternatives will be examined with consideration of different criteria simultaneously. Among all MCDM tools, BWM (best worst method) proposed by Rezaei [7, 8] as an novel MCDM technique has considerable advantages compare to others methods such as (i) the number of pairwise comparison is shortened, (ii) the obtained values are much more reliable as they would be obtained from min–max optimization problem; therefore, the results are optimum, (iii) in BWM, decision makers would only use the numbers with 1–9, (iv) BWM has better dealing with inconsistency of decision making problem using consistency ratio. Although, BWM has been widely applied in many different domains [4, 9–16]. Considering all proposed extensions of BWM by scholars, this suffers from some important shortages as the following. First of all, decisions made as result of BWM utilization cannot reflect the confidence level of decision. This means that, how much we should confidence about the importance weights obtained from BWM. Secondly, the complex decision-making problem are time dependent. BWM similar to all MCDM tools are static and would fail to present dynamic features of the

M. Yazdi (✉)
Faculty of Engineering and Applied Science, Memorial University of Newfoundland, St. John's, NL A1B 3X5, Canada

S. Adumene
School of Ocean Technology, Marine Institute Memorial University of Newfoundland, St. John's, NL, Canada

E. Zarei
Centre for Risk, Integrity and Safety Engineering (C-RISE), Faculty of Engineering and Applied Science, Memorial University of Newfoundland, St. John's, NL A1B 3X5, Canada

M. Yazdi, *Linguistic Methods Under Fuzzy Information in System Safety and Reliability Analysis*, Studies in Fuzziness and Soft Computing 414,
https://doi.org/10.1007/978-3-030-93352-4_8

decision-making problem. This means that the BWM is memory lessness. Finally, in the existing extensions of BWM, this cannot reflect those factors that change continuously in nature and uses a single value for comparison process.

Considering the aforementioned shortages as well as high flexibility of original form of BWM, the effective method is required to solve the decision-making problem, which is also the main motivation of present work. Therefore, in this study, BWM is extended with Bayesian network. To deal with the drawbacks as mentioned above, Bayesian Network is utilized in two different ways as (i) Constructing the Bayesian Network according to the prior knowledge come up from decision-makers' opinions, and (ii) learning the network using conditional probability based on the considerable input data. The first type provides lots of subjective uncertainty as this complexly up to subjective opinions from decision-makers. The probability distributions can be engaged to tackle the objective uncertainty by describing the continuous variables in Bayesian Network. Besides, the Bayesian inference process can update the Bayesian Network with both predictive and diagnostic analysis once the new evidence(s) are obtained. Considering the inherent features of Bayesian Network such as reasoning and modelling, this tool and its extensions have been extensively applied in several fields [17–25]. Bayesian Network has considerable capability to aggregate inconsistent information, quantify different uncertainties, measure dependency between the factors, and high flexibility and efficiency to make an optimum decision making [26–28].

As it can be understood from the state of the arts, Bayesian Network has been accepted as a robust decision-making tool under uncertainty. Therefore, this can be utilized to make a marginal decision to evaluate hospital service quality with consideration of confidence level. There is no study to assess hospital service quality or even more similar problems using Bayesian Network. There is no study to assess hospital service quality or even more similar problems using Bayesian Network. Since there is a complex decision-making environment in hospital quality service analysis, the contributions of this study are two folds: (i) a combination of BWM and Bayesian network is proposed to analyze the hospital service quality, and (ii) Bayesian learning is utilized to develop a much more viable and acceptable dependency for continuous bivariate distributions.

In the next section, a new probabilistic-based hybrid model (BWM-Bayesian network) is proposed and followed by an application of study. Finally, a conclusion and further remarks are provided in the last section.

8.2 Proposing a New Probabilistic-Based Hybrid Model (Fuzzy-BWM-Bayesian Network)

Step one: Defining the decision-making problem

The fuzzy-based hybrid model has demonstrated applicability for complex engineering energy systems. To further understand how the systems' complexity and non-linear interaction could propagate to performance degradation of the offshore system, a hybrid fuzzy-Bayesian network (Fuzzy-BN) is explored. Figure 8.1 shows the schematic of the algorithm of the Fuzzy-BN for offshore energy system reliability prediction. The subsections briefly describe the procedure of the model application.

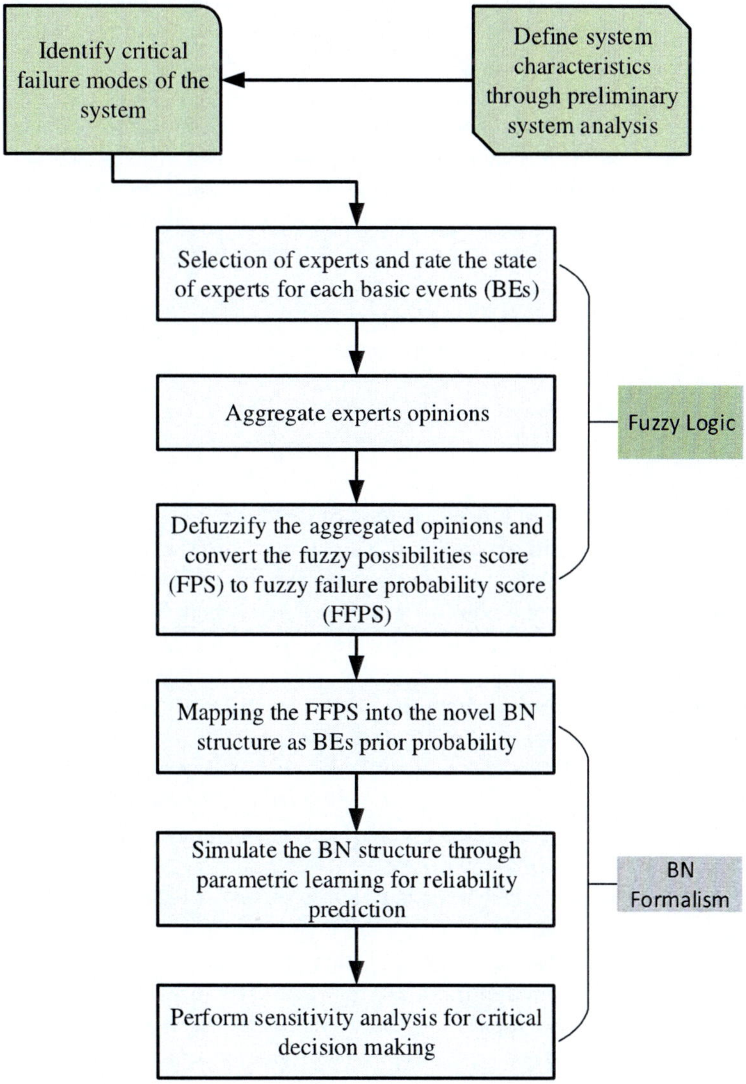

Fig. 8.1 An algorithm for hybrid Fuzzy-BN model for reliability prediction

In the proposed methodology, the state of the offshore wind system is first defined via preliminary analysis and establish the critical failure modes of the system and their interrelationship from literature and discussion are made with the domain experts involved in the operations of the offshore wind energy system. Upon the definition of the system and due to sparse operational data on the system performance, expert opinions are resorted to for the failure probability prediction. Expert opinions have been applied in modeling risk-related analysis through linguistically expressed values and descriptions. Linguistics values are adopted to characterize phenomena that are complex with multiple dependencies. These linguistic values can be formalized as membership functions of fuzzy sets, and their appropriate types (such as trapezoidal and triangular fuzzy numbers) are dependent on the system of study.

The selection of experts is made based on their period of learning, offshore field experience, and analytical behavior in the energy domain of knowledge. These criteria thus influence the experts' judgment and the specific situations of the diverse group of experts. For any given expert opinions, the application of the Delphi approach as proposed by [29], for the calculation of expert's weight. For instance, if the expert is more knowledgeable by means of experience and learning, it will be given a greater score. Equations (8.1) and (8.2) gave a simplified expression for the weighting score and factor of experts for a given experts scores of $E_1, E_2, E_3 \ldots E_n$

$$\text{Weight score of } E_i = X_{IP} \text{ score of } E_i + X_{FE} \text{ score of } E_i + X_{AQ} \text{ score of } E_i + \text{Age of } E_i \tag{8.1}$$

where X_{IP} represents the industrial position of the domain experts, X_{FE} is the field experience of the domain experts, and X_{AQ} is the academic qualification of the domain experts. This is briefly shown in Table 8.1. Furthermore, the rating of the state provides experts with flexibility in their opinions about the basic events using linguistic variables. The linguistic terms are transformed into fuzzy numbers for system modeling. Various conversion scales have been proposed; however, in the present study approach where both the performance score (X) and membership degree ($\mu(X)$) in the range of 0 and 1, were represented.

$$Weight\ factor\ of\ E_i = \frac{Weight\ score\ of\ E_i}{\left(\sum_{i=1}^{n} Weight\ score\ of\ E_i\right)} \tag{8.2}$$

The expert opinions, which function from their experience and expertise, are aggregated into a consensus to have a unified value that can be processed for the failure-based analysis. Several aggregation procedures have been proposed in the open sources: however, Eqs. (8.3)–(8.7) are used to aggregate the expert's judgement based on the trapezoidal fuzzy numbers, $\tilde{X} = (a_1, a_2, a_3, a_4)$ and $\tilde{Y} = (b_1, b_2, b_3, b_4)$, particular for the adopted criterion in this study. That is, for the degree of similarity $S_{uv}\left(\check{\delta}_u \check{\delta}_v\right)$ of the opinions $\check{\delta}_u$ and $\check{\delta}_v$ of a pair of experts E_u and E_v where $S_{uv}\left(\check{\delta}_u \check{\delta}_v\right) \in (0, 1)$, Eq. 8.3 is adopted. Equations (8.4)–(8.5) are used

Table 8.1 Weighting scores and constitution of different experts

Constitution	Classification	Score
Industrial position (IP)	Operations manager	5
	R&D manager	4
	Maintenance engineer	3
	Technical staff	2
	Other	1
Field experience (FE)	> 30 years	5
	20–29	4
	10–19	3
	6–9	2
	< 5 years	1
Academic qualification (AQ)	PhD	5
	Master	4
	Bachelor	3
	OND/HND	2
	School leaver	1
Age	> 50	5
	40–49	4
	31–39	3
	20–30	2
	< 20 years	1

for the prediction of the degree of Average Agreement (AA) for expert E_u and the degree of Relative Agreement (RA) for the same expert E_u, respectively.

$$S\left(\tilde{X}, \tilde{Y}\right) = 1 - \frac{1}{4}\sum_{i=1}^{4}|a_i - b_i| \tag{8.3}$$

$$AA(E_u) = \frac{1}{N-1}\sum_{\substack{u \neq v \\ v = 1}}^{N} S\left(\check{\delta}_u, \check{\delta}_v\right) \tag{8.4}$$

$$RA(E_u) = \frac{AA(E_u)}{\sum_{u=1}^{N} AA(E_u)} \tag{8.5}$$

Further analysis to estimate the consensus coefficient degree (CC) of expert and the expert aggregated judgement $\tilde{R}_{AG}$ based on different relaxation factors can be explored using Eqs. (8.6)–(8.7).

$$CC(E_u) = \beta \times w(E_u) + (1 - \beta) \times RA(E_u) \quad (8.6)$$

$$\tilde{R}_{AG} = CC(E_1) \times \tilde{R}_1 + CC(E_2) \times \tilde{R}_2 + \cdots + \cdots CC(E_N) \times \tilde{R}_N \quad (8.7)$$

where $\beta(0 \leq \beta \leq 1)$ describes the relaxation factor that highlights the significance of expert's weight $(w(E_u))$ over relative agreement $RA(E_u)$. The values of β characterize the degree of importance. For instance, if $\beta = 0$, a homogeneous group of experts can be assessed because no importance has been assigned to the weight of experts, while for $\beta = 1$, the consensus coefficient degree of the experts is the same as their importance weight. It is important to note that the CC measures the relative worthiness of the participating experts in the decision-making process. Therefore, an appropriate value of β influences the concordance among the importance weight and degree of relative agreement. It is the decision maker's responsibility to choose the β values based on the study scenario appropriately. $\tilde{R}_i(i = 1, 2, \ldots N)$ defines the subjective rating selected by chosen criteria relative to the experts' alternatives.

The aggregated fuzzy number generated from the experts are defuzzified. Defuzzification methodology transforms the output from the fuzzy domain back into the crisp domain to produce a quantifiable result in the fuzzy logic. For this research, the Centre area defuzzification techniques are adopted, as shown in Eq. (8.8). The predicted fuzzy possibilities values are converted into fuzzy failure probabilities to serve as input parameters for the BN structure. This transformation process captures the subjectivity nature of the data to propagate through the parametric learning of the BN. The transformation technique, as shown in Eq. (8.9), is used to convert the fuzzy possibility values to fuzzy failure probability values in this research.

$$\begin{aligned} X^* &= \frac{\int_{a_1}^{a_2} \frac{x-a}{a_2-a_1} x dx + \int_{a_2}^{a_3} x dx + \int_{a_3}^{a_4} \frac{a_4-x}{a_4-a_3} x dx}{\int_{a_1}^{a_2} \frac{x-a}{a_2-a_1} dx + \int_{a_2}^{a_3} dx + \int_{a_3}^{a_4} \frac{a_4-x}{a_4-a_3} dx} \\ &= \frac{1}{3} \frac{(a_4 + a_3)^2 - a_4 a_3 - (a_1 + a_2)^2 + a_1 a_2}{a_4 + a_3 - a_1 - a_2} \end{aligned} \quad (8.8)$$

$$FPr = \begin{cases} \frac{1}{10^K} & FPs \neq 0 \\ & FPS = 0 \end{cases} \quad (8.9)$$

where,

$$K = \left[\frac{(1 - FPs)}{FPs} \right]^{(1/3)} \times 2.301$$

Step two: Translating influence diagram into the Bayesian Network

Bayesian Network (BN) is a well-known probabilistic tool to construct conditional dependency among a set of variables. For this purpose, conditional probability tables would be used. To show the dependency of variables in the Bayesian Network, a directed acyclic graph is conducted. Thus, Bayesian Network has two main parts: (i) a graphical part representing directed acyclic graph and (ii) a probabilistic part as conditional probability table. In the visual representation of the Bayesian Network, nodes are standing for variables, and arrow connection between the nodes is standing the causality in the variables. Let us have four nodes A, B, C, and D, arrows are starting from A, B, and C, then finish with D. According to this assumption, A, B, and C are parent nodes of D, and D is denoted as a child node. The parent node here D can directly affect its child nodes A, B, and C, which will be quantified using conditional probability distributions as $P($ B|parents(D)). Keeping this introduction, the child and parent nodes can be categorized into three different class: (i) root nodes (these nodes are without parent nodes), (ii) intermediate nodes (these nodes have both parent and child nodes), and (iii) leaf nodes (the nodes without and child node). Besides, both probability of A and B as two variables can occur, when the production of A and B has given X in terms of probability. To show the operational features of BN, assuming that in a typical BN, n variables as $A_1, A_2, A_3, \ldots, A_n$, are included. In this accordance, the joint probability distribution of variables could be decomposed into the $P(A_1, A_2, A_3, \ldots, A_n) = P(A_1|A_2, A_3, \ldots, A_n) \times P(A_2|A_3, \ldots, A_n) \ldots \times P(A_{n-1}|A_n)$. To get more details related to BN, one can refer to the previous literature [30–32].

Step three: Collecting the required information (subjectively or objectively)

In this step, the input data for child nodes in the structured BN are defined. For those nodes that there is availability of data, a best fitted distribution is defined. In case of lack of data, a group of decision-makers are employed to express their opinions the child nodes in BN. The decision-makers need to express their opinions to provide a proper important weight for the factors. The preference of employing decision makers would be a heterogenous group decision-makers within the relevant background for the under study decision-making problem. In contrast, the most significant task to evaluate the best fitted distribution for the variables in child nodes is properly describing a probability distribution. To handle this issue, three different types of distributions can be considered and called as the normal distribution, Beta distribution, and log-normal distribution. To evaluate the precision of marginal distributions, the comparison tools like Akaike Information Criterion (AIC) is utilized as $\text{AIC} = -2 \times \log(\text{maxlikelihood}) + 2 \times (\text{numberofparameters})$ [30]. It can be concluded that the AIC with minimum value shows that that marginal distribution is the best. In addition, AIC should be implemented to all the variables in the network in accordance, and it is not required that all variables follow an equal marginal distribution.

Step four: Obtaining the importance weights of criteria using BWM

In this step, BWM is used to obtain the weights of criteria in the BN. This step can be significantly improved having importance weights in which better reflect the real weights. Performing BWM is used with assist excel file provided in official BWM website (www.bestworstmethod.com/software). To obtain more details regarding BWM improvement one can refer to the [29, 33, 34].

Step five: Bayesian Network analysis

In this step, first by running BN model, the quality index value of understudy hospital would be derived. By comparing the different quality index, the worst and best hospital would be obtained. Second, a dynamic feature of BN is performed to see how quality index would be changed over time. Finally, backward propagation analysis is performed in which it is working entirely on a different side from forwarding propagation analysis. The main goal is to diagnose the goal's causes in complicated system dependency. The following equation using Bayes' theorem is used to compute the posterior probability distribution of causes x_i. The distribution variation is showing that how much a cause can contribute to the consequence.

8.3 Application of Study

To show the application of the new probabilistic-based hybrid model (BWM-Bayesian network), a hospital service quality in a Metropolitan is studied. The present study's alternatives and criteria to evaluate hospital service quality are obtained from the literature [35–37] and discussing the experts of the field. The relevant structural alternatives and criteria are presented in Fig. 8.1, which is an influential diagram illustrated by the Bayesian network. As mentioned in the methodology section, the decision-making problem is first defined, which is obtaining the hospital's quality index. The second is determining the alternatives and criteria and constructing the influence diagram as depicted in Fig. 8.2. In step three, the input information for relevant alternatives (child nodes in Bayesian network) is obtained from objective data as well as subjective (i.e., employing a group of decision-makers). For example, for the node alternative "Hospital staff are neat and tidy", the percentage of how much this sentence is correct is obtained. Subsequently, the best-fitted distraction derived that is the normal distribution. This process is continued for all nodes as obtaining the best-fitted disruptions based on objective data or subjective opinions from decision-makers. For example, for the node obtained objectively, "Hospital has professional medical staff" using Fig. 8.3. For this factor, more than 90% of the data points are less than 80% and the data main focus on average values.

In step four, the employed group of decision-making determining the importance weights criteria underlying the ideal of BWM. The importance weights are regarding the criteria: Hospital equipment, Service attitude, Pharmacy and medical treatment, Professional capability, administrative policy, and Hospital sanitation and environment are 0.14, 0.07, 0.15, 0.30, 0.09, and 0.25, respectively. In step five, first, the quality index of the hospital is obtained as 0.76. This value helps us to perform propagation analysis to see which alternative has more contribution. This means that alternatives in the descending ranking need to received further intervention actions [38]. According to backward propagation analysis, the three first alternatives are (i) hospital has service staff who can solve problems immediately, (ii) there is service personnel to guide accurate clinic location in the hospital, and

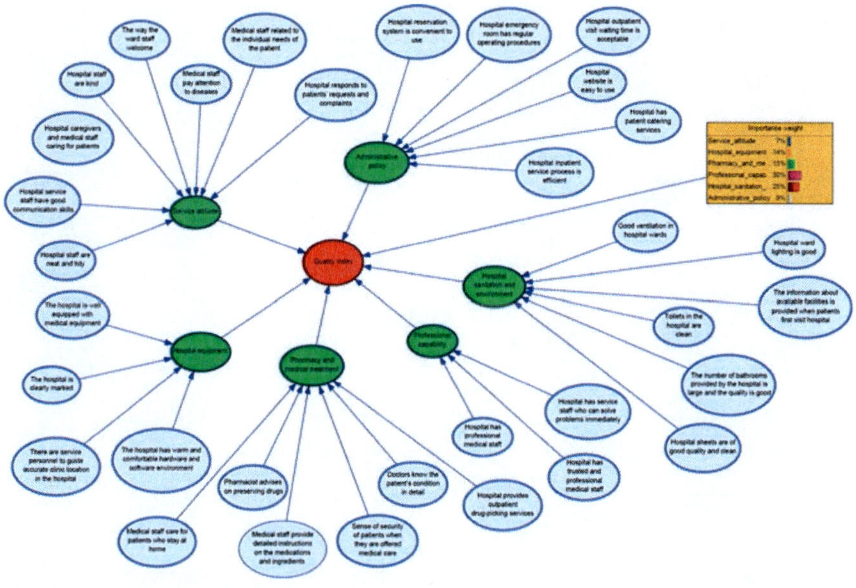

Fig. 8.2 Influential diagram to evaluate the quality index of a hospital in a Metropolitan

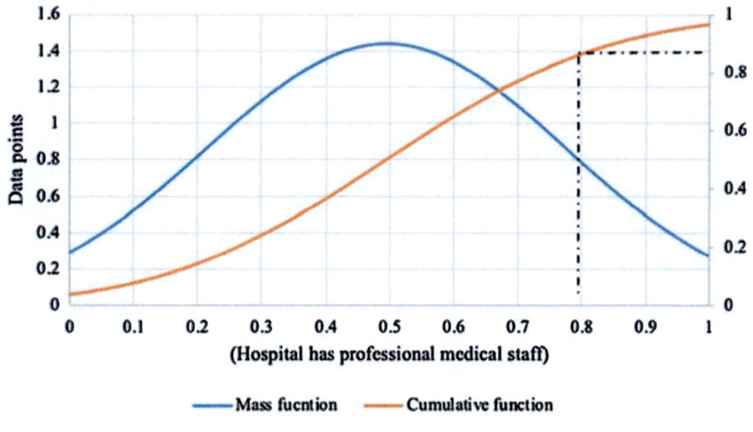

Fig. 8.3 An example of input data for variable "Hospital has professional medical staff"

(iii) hospital has professional medical staff. This ranking provides the information in which the additional intervention actions such as corrective actions can improve the hospital's quality index. Another analysis is forward propagation, in which the node "Hospital sanitation and environment" is set as the only evidence and its importance weight considered equal to be 1 (see Fig. 8.3). As it can be seen from Fig. 8.4 the nodes "Ventilation in hospital" is obtained the first ranking to receive the intervention actions. In addition, the quality index of the hospital can be assessed in a dynamic

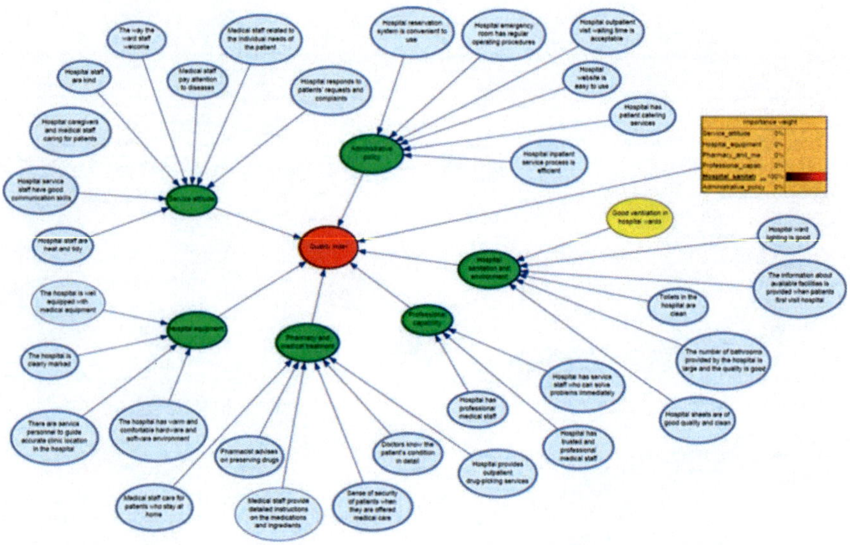

Fig. 8.4 Forward propagation analysis considering the importance weight Hospital sanitation and environment is equal to 1

manner (see Fig. 8.5). According to this point, it is assumed that the hospital would be evaluated in 10-time slices within a month based. This shows that the quality index is not memoryless and depends on the previse time slices. Updating the conational probability table in each time slice, the value quality index gradually decreases from 0.76 to 0.43 in t = 1 to t = 10, respectively. Finally, using the proposed method, quality index of different hospital can be compared. For example, four hospitals can be considered for comparison. The quality index of each hospital using BWM-Bayesian network with specific input data would be determined. Then, the quality index is ranked in the descending order.

8.4 Conclusion

This study proposed a new probabilistic-based hybrid model to solve a complex and time-dependent decision-making problem. The primary shortages of specific decision-making tools like MCDM are highlighted. The proposed method shows that it can adequately reliably deal with shortages. A hospital study application is taken into account to show the proposed method's effectiveness and efficiency compared with MCDM tools. The results show that the BWM-Bayesian network has enough capability to solve the decision-making problem over time and provide valuable information for decision-makers. For future studies, Bayesian inference can

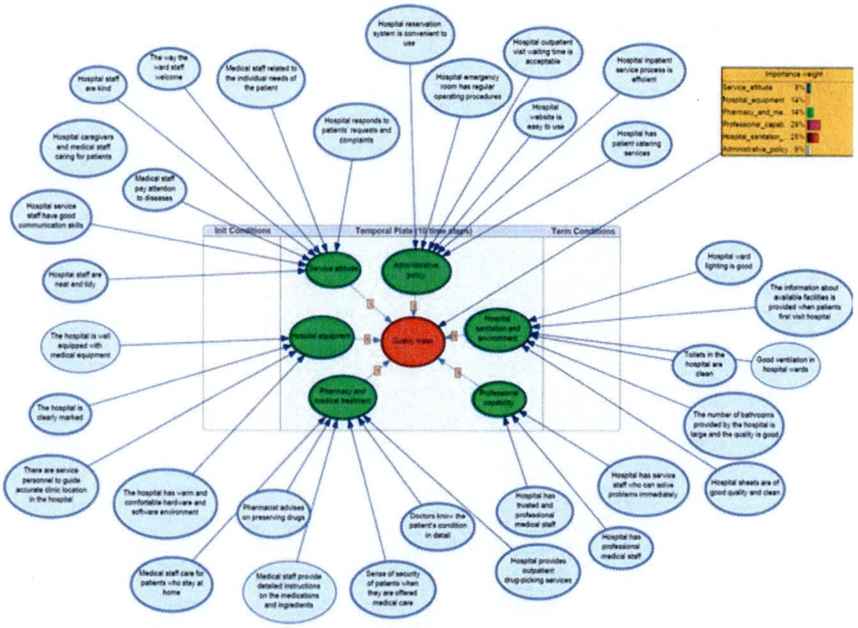

Fig. 8.5 Evaluating the quality index of the hospital in dynamic manner (10 time slices)

also be used in the Bayesian network for model improvement. Different extensions can be added to the model to better deal with subjective uncertainty.

References

1. M.C. Yu, M.H. Su, Using fuzzy DEA for green suppliers selection considering carbon footprints. Sustainability **9** (2017). https://doi.org/10.3390/su9040495
2. S.K. Lee, G. Mogi, J.W. Kim, B.J. Gim, A fuzzy analytic hierarchy process approach for assessing national competitiveness in the hydrogen technology sector. Int. J. Hydrogen Energy **33**, 6840–6848 (2008). https://doi.org/10.1016/j.ijhydene.2008.09.028
3. M. Shaverdi, I. Ramezani, R. Tahmasebi, A.A.A. Rostamy, Combining Fuzzy AHP and Fuzzy TOPSIS with financial ratios to design a novel performance evaluation model. Int. J. Fuzzy Syst. **18**, 248–262 (2016). https://doi.org/10.1007/s40815-016-0142-8
4. H.-W. Lo, J.J.H. Liou, A novel multiple-criteria decision-making-based FMEA model for risk assessment. Appl. Soft Comput. **73**, 684–696 (2018). https://doi.org/10.1016/j.asoc.2018.09.020
5. M.W. Khan, Y. Ali, F. De Felice, A. Petrillo, Occupational health and safety in construction industry in Pakistan using modified-SIRA method. Saf. Sci. **118**, 109–118 (2019). https://doi.org/10.1016/j.ssci.2019.05.001
6. R. Kaya, B. Yet, Building Bayesian networks based on DEMATEL for multiple criteria decision problems: a supplier selection case study. Expert Syst. Appl. **134**, 234–248 (2019). https://doi.org/10.1016/j.eswa.2019.05.053

7. J. Rezaei, Best-worst multi-criteria decision-making method: some properties and a linear model. Omega (United Kingdom). **64**, 126–130 (2016). https://doi.org/10.1016/j.omega.2015.12.001
8. J. Rezaei, Best-worst multi-criteria decision-making method. Omega (United Kingdom). **53**, 49–57 (2015). https://doi.org/10.1016/j.omega.2014.11.009
9. H.W. Lo, J.J.H. Liou, C.N. Huang, Y.C. Chuang, A novel failure mode and effect analysis model for machine tool risk analysis. Reliab. Eng. Syst. Saf. **183**, 173–183 (2019). https://doi.org/10.1016/j.ress.2018.11.018
10. A. Hafezalkotob, A. Hafezalkotob, A novel approach for combination of individual and group decisions based on fuzzy best-worst method. Appl. Soft Comput. J. **59**, 316–325 (2017). https://doi.org/10.1016/j.asoc.2017.05.036
11. P. Shojaei, S.A. Seyed Haeri, S. Mohammadi, Airports evaluation and ranking model using Taguchi loss function, best-worst method and VIKOR technique. J. Air Transp. Manag. **68**, 4–13 (2018). https://doi.org/10.1016/j.jairtraman.2017.05.006
12. X.F. Ding, H.C. Liu, A new approach for emergency decision-making based on zero-sum game with Pythagorean fuzzy uncertain linguistic variables. Int. J. Intell. Syst. **34**, 1667–1684 (2019). https://doi.org/10.1002/int.22113
13. A. Rostamabadi, M. Jahangiri, E. Zarei, M. Kamalinia, S. Banaee, M.R. Samaei, Model for A Novel Fuzzy Bayesian Network-HFACS (FBN-HFACS) model for analyzing Human and Organizational Factors (HOFs) in process accidents. Process Saf. Environ. Prot. **132**, 59–72 (2019). https://doi.org/10.1016/j.psep.2019.08.012
14. Ž. Stević, D. Pamučar, E.K. Zavadskas, G. Ćirović, O. Prentkovskis, The selection of wagons for the internal transport of a logistics company: a novel approach based on rough BWM and rough SAW methods. Symmetry (Basel) **9** (2017). https://doi.org/10.3390/sym9110264
15. H. Gupta, M.K. Barua, A framework to overcome barriers to green innovation in SMEs using BWM and Fuzzy TOPSIS. Sci. Total Environ. **633**, 122–139 (2018). https://doi.org/10.1016/j.scitotenv.2018.03.173
16. W. Serrai, A. Abdelli, L. Mokdad, Y. Hammal, Towards an efficient and a more accurate web service selection using MCDM methods. J. Comput. Sci. **22**, 253–267 (2017). https://doi.org/10.1016/j.jocs.2017.05.024
17. I.A. Papazoglou, O.N. Aneziris, L.J. Bellamy, B.J.M. Ale, J. Oh, Quantitative occupational risk model: single hazard. Reliab. Eng. Syst. Saf. **160**, 162–173 (2017). https://doi.org/10.1016/j.ress.2016.12.010
18. Y. Yang, F. Khan, P. Thodi, R. Abbassi, Corrosion induced failure analysis of subsea pipelines. Reliab. Eng. Syst. Saf. **159**, 214–222 (2017). https://doi.org/10.1016/j.ress.2016.11.014
19. X. Li, G. Chen, S. Jiang, R. He, C. Xu, H. Zhu, Developing a dynamic model for risk analysis under uncertainty: case of third-party damage on subsea pipelines. J. Loss Prev. Process Ind. **54**, 289–302 (2018). https://doi.org/10.1016/j.jlp.2018.05.001
20. S. Adumene, M. Okwu, M. Yazdi, M. Afenyo, R. Islam, C.U. Orji, F. Obeng, F. Goerlandt, Dynamic logistics disruption risk model for offshore supply vessel operations in Arctic waters. Marit. Transp. Res. **2**, 100039 (2021). https://doi.org/https://doi.org/10.1016/j.martra.2021.100039
21. H. Li, J.-Y. Guo, M. Yazdi, A. Nedjati, K.A. Adesina, Supportive emergency decision-making model towards sustainable development with fuzzy expert system. Neural Comput. Appl. **33**, 15619–15637 (2021). https://doi.org/10.1007/s00521-021-06183-4
22. M. Yazdi, Acquiring and sharing tacit knowledge in failure diagnosis analysis using intuitionistic and pythagorean assessments. J. Fail. Anal. Prev. **19** (2019). https://doi.org/10.1007/s11668-019-00599-w
23. M. Yazdi, A review paper to examine the validity of Bayesian network to build rational consensus in subjective probabilistic failure analysis. Int. J. Syst. Assur. Eng. Manag. **10**, 1–18 (2019). https://doi.org/10.1007/s13198-018-00757-7

24. M. Yazdi, N.A. Golilarz, A. Nedjati, K.A. Adesina, Intelligent fuzzy Pythagorean Bayesian decision making of maintenance strategy selection in offshore sectors, in *Intelligent and Fuzzy Techniques for Emerging Conditions and Digital Transformation*, eds. by C. Kahraman, S. Cebi, S. Cevik Onar, B. Oztaysi, A.C. Tolga, I.U. Sari (Springer International Publishing, Cham, 2022), pp. 598–604
25. M. Yazdi, F. Khan, R. Abbassi, Microbiologically influenced corrosion (MIC) management using Bayesian inference. Ocean Eng. (2021). https://doi.org/10.1016/j.oceaneng.2021.108852
26. M. Yazdi, F. Khan, R. Abbassi, R. Rusli, Improved DEMATEL methodology for effective safety management decision-making. Saf. Sci. **127**, 104705 (2020). https://doi.org/10.1016/j.ssci.2020.104705
27. N. Khakzad, G. Reniers, Risk-based design of process plants with regard to domino effects and land use planning. J. Hazard. Mater. **299**, 289–297 (2015). https://doi.org/10.1016/j.jhazmat.2015.06.020
28. R. He, X. Li, G. Chen, Y. Wang, S. Jiang, C. Zhi, A quantitative risk analysis model considering uncertain information. Process Saf. Environ. Prot. (2018). https://doi.org/10.1016/J.PSEP.2018.06.029
29. H. Aboutorab, M. Saberi, M.R. Asadabadi, O. Hussain, E. Chang, ZBWM: the Z-number extension of Best Worst Method and its application for supplier development. Expert Syst. Appl. **107**, 115–125 (2018). https://doi.org/10.1016/j.eswa.2018.04.015
30. N.E. Fenton, M.D. Neil, *Risk Assessment and Decision Analysis with Bayesian Networks* (2013)
31. S. Kabir, Y. Papadopoulos, Applications of Bayesian networks and Petri nets in safety, reliability, and risk assessments: a review. Saf. Sci. **115**, 154–175 (2019). https://doi.org/10.1016/j.ssci.2019.02.009
32. S. Kabir, T.K. Geok, M. Kumar, M. Yazdi, F. Hossain, A method for temporal fault tree analysis using intuitionistic fuzzy set and expert elicitation. IEEE Access **8** (2020). https://doi.org/10.1109/ACCESS.2019.2961953
33. M. Mohammadi, J. Rezaei, Bayesian best-worst method: a probabilistic group decision making model. Omega (United Kingdom), 1–8 (2019). https://doi.org/10.1016/j.omega.2019.06.001
34. A. Parhizgarsharif, A. Lork, A. Telvari, A hybrid approach based on the BWM-VIKOR and GRA for ranking facility location in construction site layout for Mehr project in Tehran. Decis. Sci. Lett. **8**, 233–248 (2019)
35. L. Li, W.C. Benton, Hospital capacity management decisions: emphasis on cost control and quality enhancement. Eur. J. Oper. Res. **146**, 596–614 (2003). https://doi.org/10.1016/S0377-2217(02)00225-4
36. L. Fei, J. Lu, Y. Feng, An extended best-worst multi-criteria decision-making method by belief functions and its applications in hospital service evaluation. Comput. Ind. Eng. **142**, 106355 (2020). https://doi.org/10.1016/j.cie.2020.106355
37. T.H. Chang, Fuzzy VIKOR method: a case study of the hospital service evaluation in Taiwan. Inf. Sci. (Ny) **271**, 196–212 (2014). https://doi.org/10.1016/j.ins.2014.02.118
38. M. Yazdi, N.A. Golilarz, A. Nedjati, K.A. Adesina, An improved lasso regression model for evaluating the efficiency of intervention actions in a system reliability analysis. Neural Comput. Appl. (2021). https://doi.org/10.1007/s00521-020-05537-8

Chapter 9
Fuzzy Linear Programming in System Safety

Mohammad Yazdi and Arman Nedjati

9.1 Introduction

Linear programming (LP) technique is the most well-known constantly operations research approach. Plenty number of scientific papers investigated numerous properties of Fuzzy linear programming problems and several innovative solution procedures for tackle with the vagueness feature of them.

A LP model represents a real-world situation as a set of constraints and objective functions including variables and parameters that are determined by the decision makers (DMs) or experts. Due to the uncertainty feature of real-world problems, non-obtainable and incomplete information, and unquantifiable data, the certain or exact values parameters are mostly hard to estimate and also the objectives or constraints might not reflect the exact conditions. Fuzzy set theory by generalization of membership function as a set is used to cope the imprecise characteristic of data. By FLP which is the result of implying fuzzy set theory to LP, decision makers could overcome such problems [1–6].

FLP, as a significant branch of operation research, is focused on the optimal assignment of restricted resources to several contending activities based on provided requirements of optimization in a fuzzy environment [7]. Firstly, Tanaka et al. [8] used the concept of fuzziness, based on the [9] theory, in mathematical programming and defined the fuzzy constraints and objective functions. Different researchers considered different types of problems such as fuzzy decision variables with crisp coefficients [10], fuzzy coefficients and fuzzy definition of feasible set [11], etc. The most cited journal in fuzzy optimization and decision making is named Fuzzy Sets and Systems, and [9] is the most cited article. The emerging keywords in this field

M. Yazdi (✉)
Faculty of Engineering and Applied Science, Memorial University of Newfoundland, St. John's, NL A1B 3X5, Canada

A. Nedjati
Industrial Engineering Department, Quchan University of Technology, Quchan, Iran

M. Yazdi, *Linguistic Methods Under Fuzzy Information in System Safety and Reliability Analysis*, Studies in Fuzziness and Soft Computing 414,
https://doi.org/10.1007/978-3-030-93352-4_9

are 'uncertain differential equation', 'uncertain analysis', and 'artificial intelligence' and the topic are mostly contributed in management and computer science [12].

Based on the improved concept of fuzzy sets developed by [9] the subjective thought of a DM is used to shape a powerful mathematical model.

A FLP optimization model could be described as follows:

$$\widetilde{\max} \tilde{Z} = \tilde{C}x$$

Subject to,

$$\tilde{A}x \cong \tilde{B}$$

$$x \geq 0$$

where $\tilde{C}$, $\tilde{A}$, and $\tilde{B}$ are fuzzy sets in R. The DM can model the real world problem based on the existing information and the feasible solutions could be described in fuzzy sets. Generally different classes of FLP models can be considered based on the objective function, constraints, or parameters coefficients type. These different types of FLP models are described in the next section.

9.2 Classification

FLP could be classified in four categories (types) as follows:

1. Fuzzy parameters LP model.
2. Fuzzy objective function LP model.
3. Fuzzy constraints LP model.
4. Fuzzy objective function and constraints LP model.

In Fuzzy parameters LP model, some parameters or all could be fuzzy numbers and the solution procedure could be categorized in two groups. The first approach is non-converted procedure in which generalizations of simplex algorithm are utilized to solve the problem. In the second approach the fuzzy LP models are converted to crisp models and solved by regular procedures. Type 2 is an LP model with fuzzy coefficients in objective function. The solution procedure of this type is suggested by [13] and some interesting facts are proved. In type 3 the resource limitations are described by membership functions, and type 4 is the complicated one which combines type 2 and 3. Verdegay [14] used the concept of variance and expectation as ranking functions to develop a method to cope with the problem of selecting a solution among alternative optimal solutions with different fuzzy objective values. However, they [15] introduced a novel method before using the concept of comparison of fuzzy numbers in FLP problems with trapezoidal fuzzy numbers in variables and RHS values. For detailed information about the models and solution approaches interested readers could refer to [16]. Maleki et al. [16] conducted that for type 1 there are two main simplex based and non-simplex based approaches, with five and two variants among with algorithms respectively, type 3 has two Verdegay and Werners approaches, and type 4 has Zimmerman and Chanas approaches.

9.3 Applications of Fuzzy Linear Programming

The applications of FLP problems varies in different fields. According to analyses results of Clarivate, the five top fields with highest number of publications involving FLP in title, abstract, or keywords are Computer Science Artificial Intelligence, Engineering Electrical Electronic, Operations Research Management Science, Computer Science Theory Methods, and Mathematics Applied respectively.

9.3.1 Computer Science Artificial Intelligence

Ebrahimnejad and Verdegay [17] studied a supplier selection problem as a well-known multi criteria decision making problem. They applied quantified SWOT integrated with triangular fuzzy numbers and fuzzy logic to tackle with the human thoughts vagueness. The decision makers are need to determine the level of suppliers by considering SWOT matrix. As a linear programming model the demands are in fuzzy numbers and the problem is solved on a real case study problem. Another cited study considers fuzzy analytic hierarchy process in new product screen application [18]. They applied linear goal programming model to normalize fuzzy weights to facilitate pairwise comparisons of fuzzy metrices. Amin et al. [19] examined interactive FLP by using Verdegay, Chanas, Werners, and Zimmermann's approaches to model the real world fuzzy in a way that interactively communicates with decision maker. The problem is applied on a textile firm problem and extensively shows the significance of fuzziness consideration in an interactive FLP and the way to apply it. The logistic outsourcing selection is another application conducted by [20] to optimize cost and improve competitiveness. The problem is modeled as a group decision making with intuitionistic fuzzy preference relations in the form of an intuitionistic FLP. The intuitionistic FLP is used to make priority weights and is solved by the pessimistic, optimistic and combined approaches. Next, TOPSIS is applied to determine the experts' weights. They applied the technique on a logistic outsourcing provider selection problem and generalized the intuitionistic fuzzy programming approach for applicable membership and -non-membership functions. Wan et al. [21] used FLP to classify the credit cardholders' behavior. Behavioral analysis is a well-known topic in credit card portfolio management. The spending history is described by the cardholder's behavior and could be evaluated by a score of combined attributes. This score is usually measured by neural networks or statistics despite some linear programming-based techniques which is extended by [21] in a heuristic based FLP to detect the bankruptcy pattern of holders. They categorizes the holder behavior in a fuzzy environment by FLP, and a real case study data from a US bank is used to prove the proposed approach. He et al. [22] is considered a FLP with fuzzy variables and parameters as a fully fuzzy problem. They developed a novel parametric technique based on alpha-cut representation of fuzzy intervals. A generic reverse logistic problem in the form of network design is proposed to show the validity and ability of making precise solutions of the approach for a risk-averse person. Baykasoılu and Subulan [23] studied a FLP problem in the energy sector. They developed a fuzzy renewable energy model to tackle with the energy problem with fuzzy parameters. The model minimizes the cost of energy producing from different energy sources with uncertain conditions. They solved the problem with exact and metaheuristic methods. A novel fuzzy linear programming model to minimize the cost of capacitated fuzzy network where the total resource limitations are fuzzy is developed and implemented to optimize the operations cost of American Airlines [24]. The study is helpful to discover the most beneficial destinations for the American Airlines. The solution of the proposed fuzzy formulation shows that flight operations capacity can be significantly increased with relatively small increase in operations cost.

9.3.2 Operations Research and Management Science

Khan and Rafique [25] designs A FLP-based technique for tactical supply chain planning in an uncertainty environment. They model the uncertainties of a multi-level, multi-product, multi-echelon supply chain network by use of fuzzy sets. The uncertain feature of the process and demand are considered jointly by the objective of minimizing cost while the customer demands are supplied. Peidro et al. [26] used a Fuzzy chance constrained linear programming model for minimizing the scrap expenses in steel production industry which is challenging since the chemical composition of scarp in uncertain in secondary steel production process. The study applies a strengthened version of soft constraints to interpret fuzziness and develop a crisp model. The Simulation results based on real world data indicate that the failure risk could be decreased by managed combination of confidence factors and spiration levels for defining fuzzy numbers. Cement transportation planning is another application of FLP models which is taken into consideration in Taiwan. Rong and Lahdelma [27] used three types of FLP methods to determine the optimal amount of transportation and the capacity of new port facilities. Their concentration is on the mathematical formulation of the problem. Another study used FLP for production planning in coal mines [28]. The problem is applied on the data from a real coal mine in Virginia, Illinois and Pennsylvania, that shows the applicability of the model for coal industry. Municipal solid-waste management is the problem which is studied by [29] as a FLP model with uncertainties in objective function and constraints. They used a direct transforming algorithm that could significantly reduce computational requirements and thus is useful for large-scale problems. The method is applied on a case of long-term waste management planning case and the results shows that solution procedure is valid. Project management is another field which is influenced by FLP models. Fan et al. [30] develops an interactive FLP approach to solve a fuzzy project management decision making problem. The objective is to minimize cost, total budget, project completion time, and the duration of jobs. They offered a systematic framework which facilitates the decision-making procedure.

9.3.3 Multi-objective FLP

As multi-objective linear programming models better reflects the real-world problems and gained more attention in pas decades, Multi-objective FLP problems are widely studied consequently. In multi-objective problems two or more contradicting objectives are determined to be optimized simultaneously, while improving one objective would weaken the other objective(s). A large number of researchers investigated on fuzzy FLP problems in different fields and interesting results are obtained. Liang [31] used bipolar point of view to model fuzzy multi-objective linear problems which traditionally solved by a max–min Zimmermann model that aggregates all fuzzy constraints and objective functions sets together. Bipolarity helps to distinguish the positive and negative preferences. The objective functions act as positive preferences while the constraints act as negative ones. Their approach handles the objectives and constrains separately and combine them in an aggregating procedure. Dubey and Mehra [32] tries to find the pareto optimal solution of a Multi-objective FLP. They assume that the decision maker considers fuzzy goals for each objective. They proved that a fuzzy solution may not be pareto optimal if one goal is fully achieved and suggested an approach to obtain fuzzy pareto optimal solutions in a general procedure. Jiménez and Bilbao [33] introduced a lexicographic order to compare arbitrary triangular fuzzy numbers to convert a fully fuzzy multi-objective LP problem to a crisp multi-level and the solution procedure. [34] represents a multi-objective FLP with fuzzy coefficients in constraints and objective functions that are interval-typed triangular fuzzy numbers. For a given level, they converted the model to a maximizing sum of membership degrees model. The point is that their membership degree are derived based on the deviation of optimal result of each objective. In the same time the constraints are changed to crisp ones by use of dominance possibility criterion. Li et al. [35] solves fully fuzzy multi-objective LP problem with fuzzy dominant degrees. He studied the relative relations within two fuzzy numbers to introduce a new measure to be able to compare two fuzzy numbers. This approach helped him address the dominant level that prefers one fuzzy number than the other. He solved the fully fuzzy FLP model by the obtained fuzzy dominant degrees and compared the results by fuzzy ranking method and indicated a better performance. In another study [36] developed a novel two-stage method for solving fuzzy FLP problems. They assumed that all resources, technological coefficients, and objective coefficients are trapezoidal fuzzy numbers. They introduced an order relationship for trapezoidal fuzzy numbers by interval expectations. In a two-stage solution procedure with the help of TOPSIS approach the problem is solved and applied on an engineering portfolio selection case.

9.4 Future Direction

Real world problems are combined with different types of uncertainties which could be classified as probabilistic or possibilistic groups. FLP is a well-known topic by very wide area of applications and the necessity of developing applicable solution procedures. Despite various types of models studied in literature, there is a significant potential for improving the algorithms and employing fuzzy concept in linear programming more efficiently. Specially in multi-objective LP problems which are more realistic based on the real world problems, fuzzy concept could be widely investigated and more reliable algorithms must be developed. The suggested future direction topics can be listed as follows:

- Multi-objective fuzzy metaheuristic algorithms
- Stochastic fuzzy linear programming models
- MCDM fuzzy approach in combination with LP
- Fuzzy LP problems in emergency situations.

However, the number of survey JCR papers are not too much in this field, some significant ones are worth to study for interested readers. Dong and Wan [37] provided a survey to a new class of fuzzy stochastic linear programming which deals with situations that fuzziness and randomness are combined in a LP framework. Luhandjula [38] conducted a survey of duality in FLP. They believe that FLP is far away from unifying theory and suggest comprehensive duality theories for specific fuzzy classes while dispensing with restrictive mathematical assumptions as a future direction. Maleki et al. [16] conducted a comprehensive survey of methods and applications of FLP and represents the FLP classification and the investigated approaches. Moreover, Schryen and Hristova [39] reviews the FLP problems with fuzzy coefficients of objective function and the solution procedures.

References

1. H. Shams, M.D. Mogouee, F. Jamali, A. Haji, A Survey on Fuzzy Linear Programming, pp. 117–133 (2012)
2. N.A. Golilarz, M. Mirmozaffari, T.A. Gashteroodkhani, L. Ali, H.A. Dolatsara, A. Boskabadi, M. Yazdi, Optimized wavelet-based satellite image de-noising with multi-population differential evolution-assisted harris hawks optimization algorithm. IEEE Access **8**, 133076–133085 (2020). https://doi.org/10.1109/ACCESS.2020.3010127

3. H. Li, J.-Y. Guo, M. Yazdi, A. Nedjati, K.A. Adesina, Supportive emergency decision-making model towards sustainable development with fuzzy expert system. Neural Comput. Appl. **33**, 15619–15637 (2021). https://doi.org/10.1007/s00521-021-06183-4
4. M.H. Memon, N.A. Golilarz, J. Li, M. Yazdi, A. Addeh, Early detection of Covid-19 disease using computed tomography images and optimized CNN-LSTM, in *2020 17th International Computer Conference on Wavelet Active Media Technology and Information Processing (ICCWAMTIP)*, pp. 161–165 (2020). https://doi.org/10.1109/ICCWAMTIP51612.2020.9317334
5. A. Nedjati, M. Yazdi, R. Abbassi, A sustainable perspective of optimal site selection of giant air-purifiers in large metropolitan areas. Environ. Dev. Sustain. (Springer, Netherlands, 2021). https://doi.org/10.1007/s10668-021-01807-0
6. M. Yazdi, A. Nedjati, R. Abbassi, Fuzzy dynamic risk-based maintenance investment optimization for offshore process facilities. J. Loss Prev. Process Ind. 194–207 (2019). https://doi.org/10.1016/j.jlp.2018.11.014
7. N. Developments, Fuzzy logic in its 50th Year, New Developments, Directions and Challenges (2016)
8. H. Tanaka, T. Okuda, K. Asai, On fuzzy-mathematical programming. J. Cybern. **3**, 37–46 (2008). https://doi.org/10.1080/01969727308545912
9. L. Zadeh, Fuzzy sets. Inf. Control. **8**, 338–353 (1965). https://doi.org/10.1109/2.53
10. H. Tanaka, K. Asai, Fuzzy solution in fuzzy linear programming problems. IEEE Trans. Syst. Man Cybern. SMC-14, 325–328 (1984). https://doi.org/10.1109/TSMC.1984.6313219
11. F. Herrera, M. Kovács, J.L. Verdegay, Optimality for fuzzified mathematical programming problems: a parametric approach. Fuzzy Sets Syst. **54**, 279–285 (1993). https://doi.org/10.1016/0165-0114(93)90373-P
12. D. Yu, Z. Xu, W. Wang, A bibliometric analysis of Fuzzy Optimization and Decision Making (2002–2017). Fuzzy Optim. Decis. Mak. **18**, 371–397 (2019). https://doi.org/10.1007/s10700-018-9301-8
13. J.L. Verdegay, A dual approach to solve the fuzzy linear programming problem. Fuzzy Sets Syst. **14**, 131–141 (1984). https://doi.org/10.1016/0165-0114(84)90096-4
14. H.R. Maleki, M. Mashinchi, Fuzzy number linear programming: a probabilistic approach (3). J. Appl. Math. Comput. **15**, 333–341 (2004). https://doi.org/10.1007/BF02935766.
15. H.R. Maleki, M. Tata, M. Mashinchi, Linear programming with fuzzy variables. Fuzzy Sets Syst. **109**, 21–33 (2000). https://doi.org/10.1016/S0165-0114(98)00066-9
16. A. Ebrahimnejad, J.L. Verdegay, A survey on models and methods for solving fuzzy linear programming problems. Stud. Fuzziness Soft Comput. **341**, 327–368 (2016). https://doi.org/10.1007/978-3-319-31093-0_15
17. S.H. Amin, J. Razmi, G. Zhang, Supplier selection and order allocation based on fuzzy SWOT analysis and fuzzy linear programming. Expert Syst. Appl. **38**, 334–342 (2011). https://doi.org/10.1016/J.ESWA.2010.06.071
18. Y.M. Wang, K.S. Chin, A linear goal programming priority method for fuzzy analytic hierarchy process and its applications in new product screening. Int. J. Approx. Reason. **49**, 451–465 (2008). https://doi.org/10.1016/J.IJAR.2008.04.004
19. I. Ertuğrul, A. Tuş, Interactive fuzzy linear programming and an application sample at a textile firm. Fuzzy Optim. Decis. Mak. **6**, 29–49 (2007). https://doi.org/10.1007/S10700-006-0023-Y
20. S.P. Wan, F. Wang, L.L. Lin, J.Y. Dong, An intuitionistic fuzzy linear programming method for logistics outsourcing provider selection. Knowledge-Based Syst. **82**, 80–94 (2015). https://doi.org/10.1016/J.KNOSYS.2015.02.027
21. J. He, X. Liu, Y. Shi, W. Xu, N. Yan, Classifications of credit cardholder behavior by using fuzzy linear programming. Int. J. Inf. Technol. Decis. Mak. **03**, 633–650 (2004). https://doi.org/10.1142/S021962200400129X
22. A. Baykasoǧlu, K. Subulan, An analysis of fully fuzzy linear programming with fuzzy decision variables through logistics network design problem. Knowledge-Based Syst. **90**, 165–184 (2015). https://doi.org/10.1016/J.KNOSYS.2015.09.020

23. S. Emeç, G. Akkaya, Developing a new optimization energy model using fuzzy linear programming. J. Intell. Fuzzy Syst. **40**, 9529–9542 (2021). https://doi.org/10.3233/JIFS-201994
24. I.U. Khan, F. Rafique, Minimum-cost capacitated fuzzy network, fuzzy linear programming formulation, and perspective data analytics to minimize the operations cost of American airlines. Soft Comput. **25**, 1411–1429 (2021). https://doi.org/10.1007/S00500-020-05228-5
25. D. Peidro, J. Mula, M. Jiménez, M. del Mar Botella, A fuzzy linear programming based approach for tactical supply chain planning in an uncertainty environment. Eur. J. Oper. Res. **205**, 65–80 (2010). https://doi.org/10.1016/J.EJOR.2009.11.031
26. A. Rong, R. Lahdelma, Fuzzy chance constrained linear programming model for optimizing the scrap charge in steel production. Eur. J. Oper. Res. **186**, 953–964 (2008). https://doi.org/10.1016/J.EJOR.2007.02.017
27. L.H. Shih, Cement transportation planning via fuzzy linear programming. Int. J. Prod. Econ. **58**, 277–287 (1999). https://doi.org/10.1016/S0925-5273(98)00206-0
28. P.C. Pendharkar, A fuzzy linear programming model for production planning in coal mines. Comput. Oper. Res. **24**, 1141–1149 (1997). https://doi.org/10.1016/S0305-0548(97)00024-5
29. Y.R. Fan, G.H. Huang, Y.P. Li, M.F. Cao, G.H. Cheng, A fuzzy linear programming approach for municipal solid-waste management under uncertainty. Eng. Optim. **41**, 1081–1101 (2009). https://doi.org/10.1080/03052150902866569
30. T.F. Liang, Project management decisions using fuzzy linear programming. Int. J. Syst. Sci. **37**, 1141–1152 (2006). https://doi.org/10.1080/00207720601014396
31. D. Dubey, A. Mehra, A bipolar approach in fuzzy multi-objective linear programming. Fuzzy Sets Syst. **246**, 127–141 (2014). https://doi.org/10.1016/J.FSS.2013.07.017
32. M. Jiménez, A. Bilbao, Pareto-optimal solutions in fuzzy multi-objective linear programming. Fuzzy Sets Syst. **160**, 2714–2721 (2009). https://doi.org/10.1016/J.FSS.2008.12.005
33. Multi-Objective Fully Fuzzy Linear Programming Problems With Triangular Fuzzy Numbers-Web of Science Core Collection (n.d.)
34. C. Li, A fuzzy multi-objective linear programming with interval-typed triangular fuzzy numbers. Open Math. **17**, 607–626 (2019). https://doi.org/10.1515/MATH-2019-0048
35. N. Van Hop, Solving fully fuzzy multi-objective linear programming problem with fuzzy dominant degrees. J. Intell. Fuzzy Syst. **39**, 3577–3595 (2020). https://doi.org/10.3233/JIFS-191888
36. J. Dong, S. Wan, A new method for solving fuzzy multi-objective linear programming problems. Iran. J. Fuzzy Syst. **16**, 145–159 (2019). https://doi.org/10.22111/IJFS.2019.4651
37. M.K. Luhandjula, Fuzzy stochastic linear programming: survey and future research directions. Eur. J. Oper. Res. **174**, 1353–1367 (2006). https://doi.org/10.1016/J.EJOR.2005.07.019
38. G. Schryen, D. Hristova, Duality in fuzzy linear programming: a survey. OR Spectr. **37**, 1–48 (2015). https://doi.org/10.1007/S00291-013-0355-2
39. H.J. Rommelfanger, Optimization of fuzzy objective functions in fuzzy (multicriteria) linear programs—a critical survey. Adv. Soft Comput. **42**, 324–333 (2007). https://doi.org/10.1007/978-3-540-72434-6_33

Chapter 10
Step Forward on How to Treat Linguistic Terms in Judgment in Failure Probability Estimation

Mohammad Yazdi and Esmaeil Zarei

10.1 Introduction

To apply the probabilistic safety and reliability analysis techniques, such as fault tree analysis (FTA), event tree analysis (ETA), Bayesian network (BN), and reliability block diagrams etc. it requires estimating the probability of an event causing the occurrence of the highest event in FTA called top event (TE) [1]. Using reliability handbook data is an option to approximate the failure probability of the events in probabilistic techniques. In some cases, when there is no information in the reliability handbook, assessors can rely on the opinions of experts which are called expert judgments. Expert judgment is a process, begin from employing experts, and collecting and aggregating their opinions. In a conventional technique, experts directly express their opinions about the probability of an event occurrence. The simple way averaging method is used to aggregate their opinions, in some cases with consideration of experts' importance weights. However, most experts prefer to express their opinions in a qualitative way. It is the point that the fuzzy set theory [2] apply to estimate failure probabilities by transferring qualitative terms into possibility and probability subsequently. Fuzzy set theory is used in safety and reliability in two different ways, the first is fuzzy inference, which is the out of the scope of this short commination, and the second is fuzzy sets, which our study is concerning using fuzzy set-in failure probability estimation.

To show how fuzzy set theory and its extensions (i.e., Intuitionistic Pythagorean, etc.) have been widely used in safety and reliability areas, a bibliometric analysis was performed concerning the number of papers published per year by the end of

M. Yazdi (✉)
Faculty of Engineering and Applied Science, Memorial University of Newfoundland, St. John's, NL A1B 3X5, Canada

E. Zarei
Centre for Risk, Integrity and Safety Engineering (C-RISE), Faculty of Engineering and Applied Science, Memorial University of Newfoundland, St. John's, NL A1B 3X5, Canada

M. Yazdi, *Linguistic Methods Under Fuzzy Information in System Safety and Reliability Analysis*, Studies in Fuzziness and Soft Computing 414,
https://doi.org/10.1007/978-3-030-93352-4_10

September 2021. The idea behind this bibliometric analysis was to make a clear distribution of both the published papers in different groups and the existing research trends in the field of safety and reliability analysis using fuzzy set theory, which can provide valuable understanding of the decision-making problems for researchers working in this field. Figure 10.1 demonstrates that safety and reliability analyses using fuzzy set theory and it's different extended methods are commonly utilized in various domains, predominantly "Engineering electrical electronic" (108 papers, 25.2%), "Computer science artificial intelligence" (77 papers, 18%), "Engineering multidisciplinary" (54 papers, 12.6%), "Computer science theory methods" (40 papers, 9.3%), "Computer science information systems" (31 papers, 7.2%), and more.

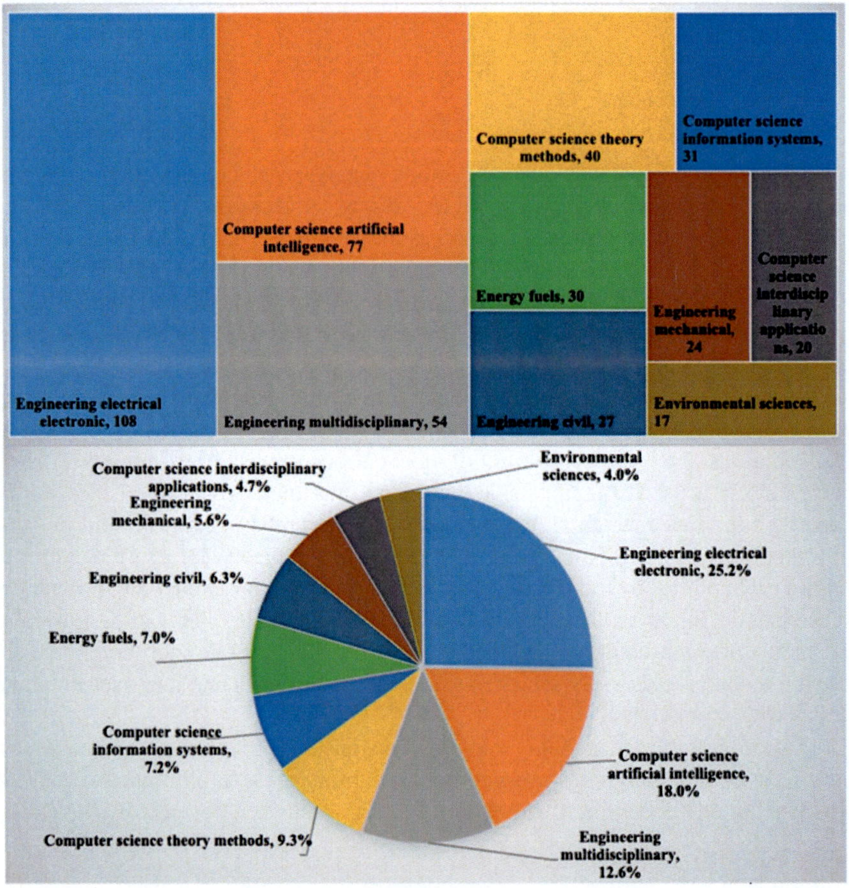

Fig. 10.1 Distribution of published works according to the different application areas by the end of September 2021 in the area of safety and reliability analysis using fuzzy set theory and its different extended techniques (*Source* Web of Science, keywords search: (Title: "fuzzy set theory" AND "reliability" OR "safety"))

Among all contributions, a couple of studies were selected to illustrate how fuzzy set theory can improve expert judgment treatment. Pasman and Rogers [3] reviewed different types of experts' judgment approaches, with details, to show how expert judgment can be treated. Authors show that the expert elicitation procedure is full of uncertainty, which further depends on the reliability of experts' opinions. Considering this point, Yazdi et al. [4] reviewed and Zarei et al. [5] uncertainty handling in probabilistic risk assessment methods showing that a combination of fuzzy set theory and BN can properly deal with the uncertainty of expert judgment. The fuzzy set theory and its extensions in probabilistic risk assessment techniques have been widely used. Some of the studies are provided as the following but are not limited to, the triangular fuzzy set is used to estimate probability in Bow-tie analysis in the gas station as an application [6]; the fuzzy AHP and fuzzy TOPSIS are used to estimate the failure probability of the events [7]. The employed experts using the Z-number method can express their opinions as well as how much they are sure about their judgment. Kumar and Kaushik [8] proposed intuitionistic fuzzy fault tree analysis (IFFTA), which is an extension to the fuzzy set theory. In another study, Kabir et al. [9] also used an intuitionistic fuzzy set to obtain the failure probability of temporal fault trees, while Zarei and co-workers integrated fuzzy logic, expert judgment, intuitionistic fuzzy AHP, fuzzy Best Worst Method, and Human Factor Analysis and Classification System (HFACS) into Fuzzy Bayesian Network to probabilistically analyze risk and human factors in the chemical process operations [10, 11]. Moreover Omidvar and et al. [12] used fuzzy hybrid multicriteria decision-making (MCDM) model for quantifying and evaluating resilience using the fuzzy Analytic hierarchy process (F-AHP) and fuzzy VIKOR (F-VIKOR) techniques under uncertainty.

The main contribution of the current study is providing critical comments on fuzzy set theory-based applications and proposing an alternative to step forward to treat expert judgment in failure probability estimation.

This paper is provided as short communication work and the organization of the paper is provided as the following. In the Sect. 10.2, the problem statement is explained with a simple process-based example. Finally, in the Sect. 10.3, the conclusion and further remarks are explained.

10.2 Problem Statement

To obtain the failure probability of an event in the probabilistic methods, assessors employed a group of experts having relevant expertise and background to the understudied system. In addition, assessors may consider the profile quality of the employed experts to consider their importance. After that, the experts express their opinions about the failure probability of the events in the provided qualitative terms. Next, assessors transferred the qualitative terms into fuzzy set numbers, which provided earlier. Aggregation methods are then used to reach unique opinions in terms of fuzzy numbers. Using the defuzzification process, a crisp possibility value would be computed. Finally, using the Onisawa equation [13], the failure probability of the events would be estimated [13–16].

There is different type of fuzzy numbers, which further cause that assessor to face different types of results. Considering chronological, one can say that there is the following development "Triangular, Trapezoidal → Intuitionistic → Pythagorean → Orthopair". Thus, using Pythagorean fuzzy numbers would provide much more reliable results compared to the as an example the triangular fuzzy numbers. That is why the rest of the fuzzy numbers are unique in nature. According to this point, the first comment is risen as "**First**: ***Different fuzzy numbers provide different probabilities, which fuzzy number is more reliable?"***. In literature, many studies use triangular fuzzy sets to estimate the failure probability of the events. What is the reason for this selection when there are more robust fuzzy sets are available? In addition, different types of aggregation methods can be used to aggregate the fuzzy numbers including but not limited to the simple averaging method, the Similarity Aggregation Method (SAM), and the power-law aggregation. Assume that there are only four types of fuzzy numbers and four types of aggregation methods. Thus, there are 16 ways to obtain the failure probability of an event. However, there are more than 16 types of fuzzy numbers and aggregation methods. The second comment rises as "**Second**: ***Different* aggregation methods *provide different probabilities. Which aggregation method is much more reliable?"*** If we step back to expert judgment many years ago, experts were free to express their opinions in probability terms from [0,1]. That is while, in fuzzy-based approaches, we are imposing the qualitative terms (set of numbers) to the experts to express their opinions like five, seven, and nine terms. Conventionally, there is an unlimited number of experts to select. The third comment is risen as "**Third**: ***Imposing qualitative terms to the experts would provide limitations to express freely their opinions?"***. Following the previous comments, imposing qualitative terms would eliminate some amount of probability in intervals between zero and one. Let us take the proposed approach by [17], if all experts express their opinions about the failure probability of a basic event as (VL (very low), VL,VL) or (VH (very high), VH, VH), the failure probabilities can be estimated as 1.63E-05 and 0.078, respectively. According to this point, the fourth comment is risen as "**Fourth**: ***By imposing qualitative terms, we are eliminating the different values of probabilities. In this case, any probability values less than 1.63E-05, and more than 0.078 are ignored and these values cannot be obtained in any combination of fuzzy numbers***. It means that the obtained probability interval is significantly smaller than usual, which is not correct.

In the following, a simple fault tree is provided to show how the probabilities obtained by different fuzzy numbers would be changed. The fire pumps are driven by a common engine. The water flow is opened by a valve. A fault tree shows the potential causes of the TE as "No water from the fire pump system" is depicted in Fig. 10.2.

It is assumed that three experts with relevant backgrounds are employed. The importance weights of experts are computed, and their opinions based on provided qualitative terms (Very low, Low, Moderate, High, and Very high) are collected. The simple averaging method is used to aggregate their opinion. Defuzzification is

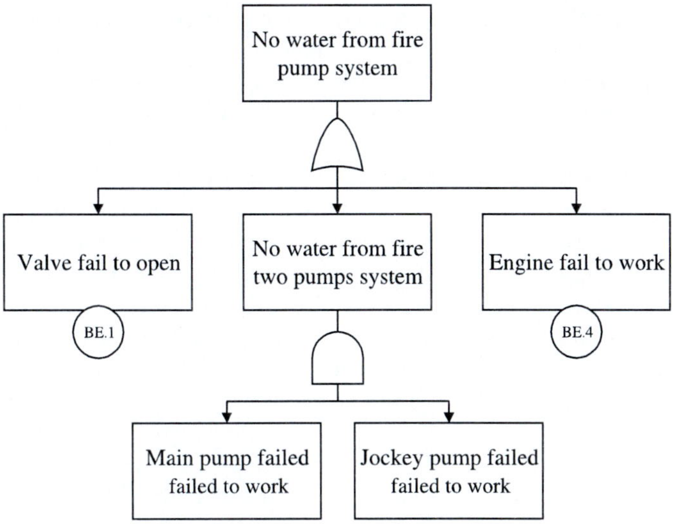

Fig. 10.2 The FTA of failure water pump

applied to obtain the crisp value of possibility. Finally, the possibly is transferred into the probability using the Onisawa formula [13, 18, 19].

In safety and reliability analysis, an important issue is improving the safety performance of the system [20]. According to this point, the probability of TE should be decreased until the safety performance of the system can be improved. In addition, the TE probability would be decreased by decreasing the probability of BEs. Using intervention actions, the failure probability will be reduced. Therefore, it is necessary to identify the critical BEs which contribute to the occurrence of TE. The critical rankings of BEs according to all type fuzzy numbers are different. Accordingly, the fifth comment is risen as "**Fifth**: ***Different fuzzy numbers provide a different critical ranking, which further brings ambiguities for assessors and subsequently possibility to make improper decision, which fuzzy number is more reliable?"***. This can be even more highlighted in the complex systems with many basic and intermediate events.

To continue, let us step back into the aggregation procedure by considering the importance weights of the employed experts. Two different elicitation procedures (the collection of experts' opinions) can be conducted to estimate failure probability. A single expert can express their opinion on the failure probability. Moreover, many independent experts can participate in an expert system problem. According to this point, two schools of thought are defined in this context. In the first one, the assessors seek the experts' opinions, called a "descriptive approach" and in the second one, the assessors, using a group of experts, find out the individual failure probability for an identified BE. This is called the "normative approach", which is commonly used in probabilistic risk assessment techniques.

To compute the importance weights, Procedure I is the individual judgment aggregation procedure. The above-mentioned sentence means that all opinions obtained from a group of experts are initially integrated into a single opinion. Accordingly, the expert system problem is treated as if there is a single expert, such as in [21–23]. In Procedure II, the importance weights for the employed experts are first derived, and accordingly, the obtained weights are combined into a single weight. This is called the individual aggregation procedure, such as in [24]. In addition, to compute the importance weight based on both different procedures, the geometric or arithmetic mean is used. It is clear the basic applications of Procedure I and II are simple; however, lots of information will be lost in such aggregation procedures. The dispersion property is completely ignored and only the centrality property is considered. Assume that in an expert system problem (group-based), if an expert expresses a dissimilar preference from the rest of exports, it will have an impressive influence on the final aggregated results. This problem would be worse if the quality profile of the experts is also considered, and as an example, the mentioned expert has a high-profile quality as well. Therefore, the sixth comment is risen as "**Sixth**: ***The aggregated results from experts' opinions considering their importance weights are so sensitive to the outliers***". Finally, there is no information about what confidence level of aggregated results is. Recently, Yazdi [25–27] proposed a framework to consider the confidence level of expert opinions. However, the methodology cannot show that the confidence level of the final results. Thus, the seventh comment rises as "**Seventh**: ***Using different fuzzy numbers cannot provide a confidence level of final results***".

10.3 Conclusion

In this paper, the challenges associated with the reliability of using the fuzzy theory and its extensions to estimate failure probabilities in the probabilistic methods of expert judgment are discussed. A simple FTA is provided to show how conventional fuzzy-based approaches are not reliable beyond our expectations, which further causes assessors to make a wrong decision. Thus, as a direction, a proper framework is needed to treat expert judgment. In this case, Markov Chain Monte Carlo an be used to obtain posterior distribution for each BE, and TE. In addition, critical ranking might be considered. In future studies, diagnosis analysis such as forward and backward analysis can be performed to obtain more data related to the studied system.

References

1. M. Rausand, S. Haugen, *Risk Assessment: Theory, Methods, and Applications* (Wiley, 2020)
2. L. Zadeh, Fuzzy sets. Inf. Control. **8**, 338–353 (1965)

3. H.J. Pasman, W.J. Rogers, How to treat expert judgment? With certainty it contains uncertainty! J. Loss Prev. Process Ind. **66**, 104200 (2020). https://doi.org/10.1016/j.jlp.2020.104200
4. M. Yazdi, S. Kabir, M. Walker, Uncertainty handling in fault tree based risk assessment: state of the art and future perspectives. Process Saf. Environ. Prot. **131**, 89–104 (2019). https://doi.org/10.1016/j.psep.2019.09.003
5. E. Zarei, N. Khakzad, V. Cozzani, G. Reniers, Safety analysis of process systems using Fuzzy Bayesian Network (FBN). J. Loss Prev. Process Ind. **57**, 7–16 (2019). https://doi.org/10.1016/j.jlp.2018.10.011
6. E. Zarei, M. Yazdi, N. Khakzad, G. Reniers, Safety assessment of process systems using fuzzy extended bow tie (FEBT) model. Chem. Eng. Trans. **77**, 1027–1032 (2019). https://doi.org/10.3303/CET1977172
7. M. Yazdi, O. Korhan, S. Daneshvar, Application of fuzzy fault tree analysis based on modified fuzzy AHP and fuzzy TOPSIS for fire and explosion in the process industry. Int J. Occup. Saf. Ergon. **26**(2), 319–335, (2020). https://doi.org/10.1080/10803548.2018.1454636
8. M. Kumar, M. Kaushik, System failure probability evaluation using fault tree analysis and expert opinions in intuitionistic fuzzy environment. J. Loss Prev. Process Ind., 104236 (2020). https://doi.org/10.1016/j.jlp.2020.104236
9. S. Kabir, T.A.N.K.I.M. Geok, M. Kumar, M. Yazdi, F. Hossain, A method for temporal fault tree analysis using intuitionistic fuzzy set and expert elicitation. IEEE Access. **8**, 980–996 (2020). https://doi.org/10.1109/ACCESS.2019.2961953
10. A. Rostamabadi, M. Jahangiri, E. Zarei, M. Kamalinia, S. Banaee, M.R. Samaei, Model for a novel Fuzzy Bayesian Network-HFACS (FBN-HFACS) model for analyzing Human and Organizational Factors (HOFs) in process accidents. Process Saf. Environ. Prot. **132**, 59–72 (2019). https://doi.org/10.1016/j.psep.2019.08.012
11. A. Rostamabadi, M. Jahangiri, E. Zarei, M. Kamalinia, M. Alimohammadlou, A novel Fuzzy Bayesian Network approach for safety analysis of process systems; an application of HFACS and SHIPP methodology. J. Clean. Prod. **244**, 118761 (2020). https://doi.org/10.1016/j.jclepro.2019.118761
12. E. Zarei, B. Ramavandi, A.H. Darabi, M. Omidvar, A framework for resilience assessment in process systems using a fuzzy hybrid MCDM model. J. Loss Prev. Process Ind. **69**, 104375 (2021). https://doi.org/10.1016/j.jlp.2020.104375
13. T. Onisawa, A representation of human reliability using fuzzy concepts. Inf. Sci. (Ny) **45**, 153–173 (1988). https://doi.org/10.1016/0020-0255(88)90038-2
14. M. Yazdi, S. Kabir, A fuzzy Bayesian network approach for risk analysis in process industries. Process Saf. Environ. Prot., **111** (2017). https://doi.org/10.1016/j.psep.2017.08.015
15. S. Kabir, M. Yazdi, J.I. Aizpurua, Y. Papadopoulos, uncertainty-aware dynamic reliability analysis framework for complex systems. IEEE Access., **6** (2018). https://doi.org/10.1109/ACCESS.2018.2843166.
16. T. Onisawa, Subjective analysis of system reliability and its analyzer. Fuzzy Sets Syst. **83**, 249–269 (1996). https://doi.org/10.1016/0165-0114(95)00381-9
17. M. Yazdi, Hybrid probabilistic risk assessment using fuzzy FTA and fuzzy AHP in a process industry. J. Fail. Anal. Prev. **17** (2017). https://doi.org/10.1007/s11668-017-0305-4
18. T. Onisawa, An application of fuzzy concepts to modelling of reliability analysis. Fuzzy Sets Syst. **37**, 267–286 (1990). https://doi.org/10.1016/0165-0114(90)90026-3
19. M. Yazdi, Footprint of knowledge acquisition improvement in failure diagnosis analysis. Qual. Reliab. Eng. Int. **35** (2019). https://doi.org/10.1002/qre.2408
20. ISO 31000: Risk Management, Int. Stand. Organ. (2009)
21. B. Blagojevic, B. Srdjevic, Z. Srdjevic, T. Zoranovic, Heuristic aggregation of individual judgments in AHP group decision making using simulated annealing algorithm. Inf. Sci. (Ny) **330**, 260–273 (2016). https://doi.org/10.1016/j.ins.2015.10.033
22. M. Yazdi, N.A. Golilarz, A. Nedjati et al., An improved lasso regression model for evaluating the efficiency of intervention actions in a system reliability analysis. Neural. Comput. Appl. **33**, 7913–7928 (2021). https://doi.org/10.1007/s00521-020-05537-8

23. M. Yazdi, T. Saner, M. Darvishmotevali, Application of an artificial intelligence decision-making method for the selection of maintenance strategy, eds. by R. Aliev, J. Kacprzyk, W. Pedrycz, M. Jamshidi, M. Babanli, F. Sadikoglu, in *10th International Conference on Theory and Application of Soft Computing, Computing with Words and Perceptions - ICSCCW 2019. ICSCCW 2019. Advances in Intelligent Systems and Computing*, vol. 1095. (Springer, Cham, 2020). https://doi.org/10.1007/978-3-030-35249-3_31
24. I. Vinogradova, V. Podvezko, E.K. Zavadskas, The recalculation of the weights of criteria in MCDM methods using the Bayes approach. Symmetry (Basel). **10**, 1–18 (2018). https://doi.org/10.3390/sym10060205
25. M. Yazdi, Introducing a heuristic approach to enhance the reliability of system safety assessment. Qual. Reliab. Eng. Int. **35** (2019). https://doi.org/10.1002/qre.2545
26. M. Yazdi, A perceptual computing–based method to prioritize intervention actions in the probabilistic risk assessment techniques. Qual. Reliab. Eng. Int. **36**, 187–213 (2020). https://doi.org/10.1002/qre.2566
27. M. Yazdi, Ignorance-aware safety and reliability analysis: a heuristic approach. Qual. Reliab. Eng. Int. **36**, 652–674 (2019). https://doi.org/10.1002/qre.2597